Agroecological Economics

Agroecological Economics

Sustainability and Biodiversity

Paul A. Wojtkowski

AMSTERDAM • BOSTON • HEIDELBERG • LONDON • NEW YORK • OXFORD
PARIS • SAN DIEGO • SAN FRANCISCO • SINGAPORE • SYDNEY • TOKYO
Academic Press is an imprint of Elsevier

Academic Press is an imprint of Elsevier
84 Theobald's Road, London WC1X 8RR, UK
Radarweg 29, PO Box 211, 1000 AE Amsterdam, The Netherlands
30 Corporate Drive, Suite 400, Burlington, MA 01803, USA
525 B Street, Suite 1900, San Diego, CA 92101-4495, USA

First edition 2008

British Library Cataloguing-in-Publication Data
A catalogue record for this book is available from the British Library

Library of Congress Cataloguing-in-Publication Data
A catalogue record for this book is available from the Library of Congress

ISBN: 978-0-12-374117-2

For information on all Academic Press publications visit our web site at www.books.elsevier.com

Typeset by Charon Tec Ltd (A Macmillan Company), Chennai, India
www.charontec.com

Printed and bound in the United States of America

08 09 10 11 10 9 8 7 6 5 4 3 2 1

All nature is but art, unknown to thee:
all chance, direction, which thou can it not see:
all discord, harmony not understood
all partial evil, universal good
Alexander Pope, *Essay on Man*

Money come, money go, money have no home
heard by this author in an African marketplace

Contents

15 Analytical Refinements 227

16 Summary 249

Preface

The conventional approach to agricultural or what also might be termed the green revolution model, i.e., monocultures of genetically improved crops supplied with ample agrochemical inputs, has, as an ideal, the image of endless fields of ripening grains. In many eyes, this signifies a forthcoming plenitude and a dominion over nature.

Not as utopian as the image projects, ecological shortcomings abound. These include the loss of natural habitats and in-residing native flora and fauna.

Additionally, nature has proven less subservient than suggested. Unabated winds can flatten unprotected grain fields, crop-eating insects and plant destroying diseases can thrive in large-scale monocrops, high-volume harvests exhaust soils, and the post-harvest situation, fields, often lacking a protective cover, expose the land to the forces of erosion. When magnified through large-scale farming, such shortcomings can have a broad and undesired impact.

Techniques and new directions have been proposed to overcome the environmental shortfalls. Born of a desire for differentiated alternatives, non-mainstream labeling can carry political and policy baggage. A case in point, the term organic farming is more an expression of principle than clearly demarcated field practices. Other labels are also weak in applied meaning.

If agroecology is to serve as an umbrella discipline, it should come without predilections or preconceived notions. A few have crept in. It should be noted that agroecology is not exclusively the control of herbivore insects without synthetic chemicals, not uniquely the application of natural compost in a backyard garden, nor is it solely within the realm of the organic producer.

Behind and linking the field-ready expressions of ecological and environmental concern, there lurks a larger field of study. At the minimum, there is the imperative that agriculture and farms, while being productive, also present a nature-friendly face. Less noted, but of concern, agroecology should have social and cultural meaning, i.e., how people interact, through agriculture, with the terrestrial world that surrounds them.

In seeking alternatives, there is bound to be an associated idealism. A lot can and is postulated, not all is attainable, at least in the form first envisioned.

Global warming cannot be stopped through agroecology, but the long-term, negative effects on agriculture can be mitigated. For other concerns, the solutions are more immediate. The lack of clean water, a crisis in far too many regions, may be remedied through improved farming practice. The negatives associated with fertilizer and other chemical runoff, including polluted water and dead zones in oceans, are arrested when agrochemicals are better contained or no longer applied. Taken together, the ideals are nice goals and undeniable byproducts of appropriate agroecology.

In tackling the larger issues, there will be disappointments and shortfalls when compared with what could be. Nevertheless, these shortfalls can still produce spectacular or, at the least, acceptable environmental results. There is nothing amiss in correcting environmental lapses but, in moving in this direction, why not seek economic benefits.

Agroecology starts at the individual farms, to the benefit of farmers, farm families, and ultimately the consumers of farm products. To be effective, the astute agroecological economist must consider many aspects. Profitability is only one. There are the hidden economics. Cropping decisions should take into account risk, native plants and animals, and societal and cultural values. They are also responsible for finding those expressions of agroecology that best fit the biology, agrology, agronomy, and natural ecology of farms and farm landscapes.

From these and a lot more, meaningful economic choices are made. It is at this level that agroecology downturns or rises. With the conviction that the latter will prove true, the thrust of this book is along economically inclusive, but targeted lines.

About the Author

As a leading proponent and analyst, Dr. Paul Wojtkowski continues to layout a vision of agroecology could be; both as an academic discipline and in how agriculture is practiced. His six previous books have affirmed the underlying motives, theories, and concepts. They have also proposed a large tally of quintessentially nature-friendly, farming practices. Although these efforts are deep in outlook, e.g., encompassing agriculture, forestry, and agroforestry, and broad in geographic scope, more insight is needed.

Economics not only expresses important differences between human-directed agroecology and natural ecology, it also holds key acceptance standards. Having observed agriculture in six continents and over 70 countries, Dr. Wojtkowski has seen what works and what doesn't. As a trained economist with advanced degrees in both agricultural and forest economics, he is able to take the next step; that of presenting agroecology as a fully-fledged science complete with its own economic underpinnings.

1 Introduction

Agroecology, an abbreviation of the term agricultural ecology, carries with it certain views. These hold that agriculture is to be studied from an ecological perspective and practiced with an ecological mandate. This mandate embodies a deference toward all things natural, including do-no-harm admonitions with regard to native flora and fauna and natural ecosystems.

These are noble goals, goals that strike a cord through the growing realization that ill-managed farms do cause environmental problems. Given the increased acquiescence that agricultural crops can be raised profitably and in an ecologically sustainable manner, agroecology has become a road worth traveling.

Irrespective of the environmental promise, there has been comparatively little interest in agroecological economics.[1] Nonetheless, in adopting biodiversity-based agroecology as the guiding format, plot-level economic questions must be asked and farmers will call for operational efficiency. This is a tall order as embracing bio-amended agriculture is, in many aspects, like staring anew. There must be a revitalized focus on long bypassed plot yield-and-cost themes. Other basic agricultural economic questions, such as plant spacing and rotational gains, once thought resolved, are again unsettled topics.

Even more pressing are how agrosystems are framed and management inputs employed to achieve environmentally sound solutions. Some of this relates to how and when synthetic chemicals or genetically engineered crops should be used, or if they should be used at all. Although agrochemical and genetic questions front the much of the immediate debate, agroecology is much more. By extension, so is agroecological economics.

ECOLOGY AND AGROECOLOGY

There is a tendency to equate ecology and agroecology. In this, there is some truth. In addition to shared ideals, the concepts and theories that underlie and explain natural ecology can and do aid in mastering agroecology.

From an economic perspective, the ecology–agroecology relationship is mostly non-existent. The reasoning being that natural ecosystems do just fine without human intervention. Under the premise that nature cannot be improved upon, except where mankind has inflicted preternatural damage, there is no need for economic opinions nor for economic input.

Although natural ecology and agroecology share a common theoretical core, the divergence is sudden and severe, so much so that these may seem separate, not interlinked, disciples.[2] For one, economics enters the picture in a big way. The production of food, fiber, and fuel must be accomplished with land use, input, and other forms of efficiency.

One obstacle to applied agroecology lies in the expanded complexity and a vastly enlarged array of options.[3] Ecological theory is helpful and can, in scattered circumstances, provide an in-depth understanding. Even within the ecology–agroecology overlap, the sorting mechanisms, those that provide meaningful economic analysis on agricultural practices, do not come from natural ecology. These remain unique to the agricultural version.

A PHILOSOPHICAL POSITIONING

In looking at the broad picture, that of the positioning of agroecology *via-a-vis* agriculture, a number of views are possible. Presented diagrammatically in Figure 1.1, each presents, in its own right, differing ranges of alternatives.[4]

There is the notion that agroecology is a subset of agriculture. This view is commonplace. Terms, such as organic gardening, permaculture, and intercropping, describe lesser agroecological subdivisions.

Some look at agroecology as a parallel discipline. Through this, agroecology addresses many of the same questions, but from an implied natural perspective. The expectation is that answers will be environmentally appropriate and ecologically sympathetic. Profit maximization does not always reign supreme, as long as there is an acceptable economic outcome, environmental mandates will force change.

There is the more inclusive view. Some look at agroecology as the umbrella discipline of which agriculture is a subset. Being this inclusive implies agroecology as a full science.

Sciences come complete with theories, methods, systematized facts, and concurring observations. An argument can be made that agroecology, with its solid body of thoughts, theories, principles, and concepts, meets this standard. The test lies in showing that the ideas that

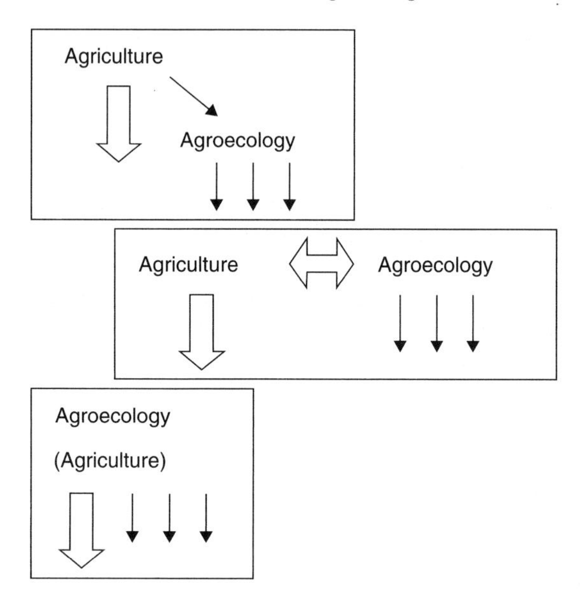

FIGURE 1.1. Three pictorial interpretations of the agriculture–agroecology relationship: (top) agriculture as the overseeing discipline with agroecology as a subset; (middle) agriculture and agroecology are separate but equal, each governing a distinct portion of the field; and (bottom) agroecology provides the all theoretical and applied oversight.

underlie agroecology are all encompassing and flow seamlessly across the totality of the science.[5]

Sciences carry other burdens. One is to link with other disciplines, e.g., sociology, anthropology, etc., further expanding the horizons. In point, cultural agroecology, i.e., the link between peoples, cultures, and agricultural practice, becomes a valid field of study.[6]

Under this stronger rendering, the economics of agroecology must transcend the profit motive to include other issues and other concerns. As a form of ecology, it should still comport strong environmental tendencies.

The agriculture/ agroecology debate is most nettlesome when a high degree of inclusiveness is sought. This text presents an inclusive view. The perspective notwithstanding, the immediate task is to provide practical answers to field-level questions. This is an excellent starting point.[7]

AGROECOLOGICAL ABSTRACTIONS

In delving into agroecological practice, differences in thought take hold. The prevailing agronomic response to threatening events, e.g., the arrival of plant-eating insects, is not to loose hard-gained productivity, but instead, to add inputs. Insecticides may be applied to keep insects out or to eliminate all when slight populations of bad insects are detected.

This situation is one of many. Carrying this further, there is the notion that agricultural systems should be productively steadfast, fighting off any and all arriving threats, e.g., droughts, plant diseases, high winds, etc. Commonly, this is done through genetically armored crops supplied with correspondingly high levels of inputs. At times, this has unfolded as a warring against nature, through inputs, to obtain high yields. This high-input, high-output version, if well formulated, need not contradict natural precepts.

The agroecological response is disparate in concept and outcome. With natural controls, a contained level of crop loss may be a desirable thing. Predator insects, those that dine on crop-eating bugs, are best entertained if they have something to consume. The complete elimination of crop-harming insects can eliminate the predator types, putting crops at a greater overall risk.[8] In agroecology, the systems should have considerable give, bending, but not breaking, as the negative forces of nature flow through.

The agroecological ideal is one where positive forces are harnessed for productive purpose. Whether this means replacing non-natural inputs, addressing weather-related threats, or surmounting other productive obstacles, nature offers an array of solutions. Although the economics remain unresolved, it is this notion, that of utilizing natural dynamics for productive purposes, that drives agroecology.

In this brief preview as to what awaits, the journey can be as rewarding as the destination. Along the way, certain assumptions are challenged, some long held beliefs discarded. Part of this lies in convincing students and practitioners that contemporary practices, good or bad, are not the only way to do things. In addition to dismissing unsound assumptions and beliefs, expanding upon the agroecological possibilities may require overcoming a complacent inertia.

TRADITIONAL SUBDIVISIONS

Agroecology is often informally subdivided. To name a few, the existent categories include:

- Conventional agronomy (mostly monocropping)
- Organic gardening[9]
- Permaculture
- Regenerative agriculture
- Low-input agriculture

- Intercropping
- Kyusei nature farming

These groupings, singular or together, often front agroecology. This is far from what should be; these subdivisions often lack precise bounds and application lucidity.[10] In some ways, these labels are counterproductive, serving more to obscure, rather than clarify, the broader picture.

Agroecology, economics included, needs to operate from a staunch, inclusive framework. Broad cross-comparisons are difficult, if not impossible, without a framework from which to analogize.[11]

For those land-use disciplines where yields or outputs are expected, agroecology provides the theoretical and conceptional core. There is a solid body of thought, theories, principles, and application concepts which support this productive process. In turn, these subdivide agroecology in ways very different than the above categories propound.

In formulating the larger picture, agriculture and agroforestry, where agricultural crops are the principle output, are clearly inclusive. Less recognized are those agroforestry and forestry practices where trees and wood are main outputs.[12] These are often judged separately as the common practical expressions, trees and wood outputs, along with the methods of management and measurement, differ from those found in agriculture. Although the practices differ, forestry and agroforestry do have, in common with agriculture, most of the underlying principles and concepts.

Within this agriculture–forestry–agroforestry context, the reach is very wide. This can be with high-intensity systems, where output and maximum land utilization is paramount, or very low-intensity practices, those where locals take, on a very limited scale, from natural ecosystems. The least intrusive end of this scale includes hunter–gatherer activities in natural forests.

THE ECONOMIC SCOPE

Redefining and expanding of the agriculture/ agroecology relationship applies to the counsel economics provide. Although monetary profit and loss of plots and farm landscapes remains a critical component, much of the economic decision process resides with cross-agroecosystem and cross-landscape comparisons.

For any given plot, there are many choices regarding the design options, i.e., the crops (one or more), accompanying biodiversity (if any), the spatial pattern and dimensions, the temporal dynamics, the types and amounts of inputs, and the various other treatments and threat regimens.

PHOTO 1.1. Traditional agriculture as expressed through high-input, season monocropping (top). In contrast, a long-term, tree/forage grass mix offers multiple outputs, higher potential profits, and an improved environmental presence (bottom).

These produce a vast array of possibilities. Comparison determines the best use of any one plot or the best designs for a plot-filled farm landscape.

This represents a shift from profit and loss as determined through monetary units on to comparison-based economics. The need for plot-comparative economics is dictated in part by the philosophical, more so by the practicalities.[13]

In having to coax yields from reluctant soils, inhospitable climates, and threat-laden surroundings, peoples in various regions have developed an array of interesting agroecological options. Integral to agroecology is risk reduction. This may be forceful enough that compromises, in the form of less chancy productivity, often hold sway over large, but unsure, monetary returns.

A degree of risk reduction is inherent with cropping biodiversity.[14] Not the only option, the techniques for reducing risk are numerous although many remain outside the agricultural mainstream.[15]

In expanding the economic scope, good agroecology should afford a pleasant local climate, an esthetically pleasing landscape, and a host of similar quality-of-life returns. The quality issues overlap with the environmental mandates. One aspect is in not misusing or misplacing potentially harmful manures, agrochemicals, and the like. In a demonstration of this, well-intentioned agroecology ensures that clean water flows from farm landscapes.

The benefits of agroecology should include making farm landscapes harmonious with natural flora and fauna. This brings on another level of analysis, one beyond profits, risk, culture and society, and quality of life, one concerned with how native flora and fauna fare in an agricultural setting. This requires effort, i.e., farms and agroecosystems must be so designed.

There are other dimensions to agroecology, those manifested through cultural, societal, and even religious values. This influence should not be ignored; it is the beliefs that people hold which dictate the types of agroecosystems people are willing to accept.

KEY ABSTRACTIONS

With so much in play, there is an overriding agroecological axiom; there is no one-best cropping solution, no one best way to do things. Any of the variations on agriculture and agroecology may produce an acceptable, or better outcome. Employing more of the agroecological options does not detract, but may help. This statement is especially credible as landusers operate in an uncertain world, with changing market prices, varying weather patterns, shifting soil characteristics, and a host of other ambiguities.

Along the way, another jump is required (at this stage, this jump is more a leap of faith). The ecosystem (plot) selection process should go beyond simple cross-ecosystem comparison on into the realm of optimization. Going back to the above axiom, this brings on the possibility, and likelihood, that many possible directions and options will produce the same global optimal.

In dealing with the added complexities, the theories, principles, and concepts of agroecology allow economists to grasp the scope and the options presented. This should not be done at an arms-length distance, i.e., economists should not look only at the results. In order to extract meaningful, broadly appertain conclusions, probing economic analysis must be mindful of the underlying agroecology.[16]

MEASUREMENT

Agroecological economics offers an array of more or less standardized methodologies and some unique to this version of economics. All have application, some more than others. Some give a narrow, focused solutions, others allow for wider, cross-ecosystem comparisons. As applied to agroecology, Figure 1.2 lays out, in schematic form, the different approaches and associated economic methodologies.

Financial analysis, that which requires monetary units for numeration, is an important end-use tool. Before going down this road, it should be noted that financial analysis is often proceeded by, or runs parallel with, comparison through efficiency units. Unique to agroecology, efficiency establishes whether or not the ecology of an agroecosystem is

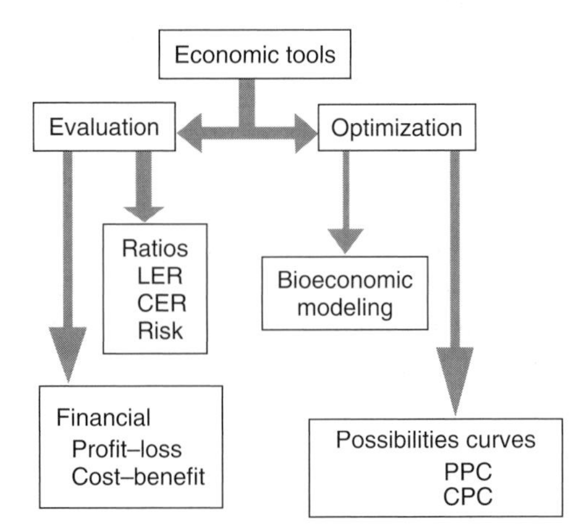

FIGURE 1.2. A listing of the economic tools of agroecology. The two main methodology sections deal with (1) plot evaluation, either financial or through efficiency ratios or (2) seek an optimized system outcome, either through possibilities curves or biomodeling.

at peak ecological and productive potency and, if not, what additional might be expected. Efficiency ratios and financial analysis are part of system evaluation.

Another methodology branch involves agroecosystem and farm optimization. Optimization can be administered through narrowly focused possibility curves or broader-based bioeconomic modeling. Restricted by severe data shortfalls, this branch is problematic. Still, there is much to be learned going, as far as possible, down even inconclusive roads.

Intangibles

In a departure from the conventional, much of economic agroecology defies quantitative analysis.[17] Instead of relying totally upon fully measurable determinants, decisions often and should be made on what some call elements of faith. In this less exacting line in inquiry, there are four forms of decision criteria:

1. Tangible and quantifiable
2. Tangible and non-quantifiable
3. Non-tangible
4. Extraneous variables

For the first of these, ratios, risk and otherwise, provide a numeric comparison. For the latter three, the applications are not as mathematically fast.

For decision variables that are tangible and non-quantifiable, acceptance that a positive interaction occurs may or may not be sufficient for consideration and inclusion into the decision process. To cite one case, hydraulic lift is a proven mechanism in polycultures. This is where one plant conveys water up from a deep source, the moisture is then transferred to a companion species.[18] Lacking a quantifiable base, the landuser must make the decision as to how far to go when putting this into practice.

Many non-tangibles are beyond any meaningful financial or economic measure. Esthetics (e.g., flowers, beautiful sunsets, etc.), comfort (e.g., a cool shady work environment), healthy surroundings (e.g., clean drinking water), or satisfaction (e.g., a pleasantly-scented cool evening) are clear-cut despite being quantitatively non-expressible. Much is lost in not considering these.

The final group, the extraneous variables, subdivide into social or technical limits. Many of these decision variables can be rightly referred to as the deadly details. These include practices that are not acceptable within a society or culture (e.g., the raising of pigs where this animal has religious taboos). Some are essentially yes-and-no criteria on whether the proposed system or change is possible.

Intuitiveness

Given the number of decision variables that underlies agroecological success, intuitiveness enters the picture in a big way. Intuitiveness is the condition where gains or losses are obvious, but this may or may not be quantifiable. With this, the first three of the above variables can be re-expressed into four intuitive–calculation and, ultimately, decision scenarios:

1. intuitive and calculable = a highly informative decision variable,
2. intuitive and incalculable = conceptionally certain,
3. unintuitive and calculable = of marginal or dubious value,
4. unintuitive and incalculable = mostly worthless for decision purposes.

Whether calculable or incalculable, if a gain is intuitive, there is a degree of certainty as a decision determinate. If unintuitive, seeming to defy logic or ecological postulation, the worth is unclear, even if accompanied by data.[19] This places it outside any immediate decision process.

It should be noted that not all agroecology is documented. A few farmers, through their own observation and volition, do go outside the pale. This can be to harness natural dynamics in unique manner or find unusual ways to better operate in inhospitable climes.[20]

Although all possible avenues are not fully interpreted, the astute agroecologist tries to operate under intuitiveness. This statement may seem abstract, but is merit laden. The starting point is the biology and ecology, proceeding on into the economics.

Through monetarily phrased units, conventional agricultural economics is often short on down-to-earth insight. Although bottom-line profit or loss decisions do coerce outcomes, users do resist, preferring to consider a wide range of economic and non-economic factors in making agroecosystem determinations.

For this, ratios are first in an arsenal of agroecological economics tools. As standards of comparison, ratios need few, if any, qualifiers. Offering values that easily compare, and knowing how these are arrived at, opens the decision process and affords intuitiveness. This facilitates the transcending step, a decision process based not upon a single number, but upon many considerations and many factors.

ENDNOTES

1. The statement on the comparative lack of interest in agroecology, at least by the economic mainstream, is verifiable through a key word search under agroecology. Few in number in the mainstream publications, more are found at the economic fringe. Between

1992 and 2002, about 500 socioeconomic articles where published in the agroforestry literature (Montambault and Alavalapati, 2005). Others are scattered in agricultural journals.

2. Vandermeer (1989), in his prefacing remarks, mentions the divergence between ecological and agroecological theory. One observation was that agroecological theory may have more application to ecology than the reverse. There is considerable truth in this.

3. The complexity of agroecology, as a barrier to development and use, is mentioned by Levins and Vandermeer (1990).

4. For a more in-depth discussion of agroecological history, definitions, etc., see Dalgaard *et al.* (2003).

5. In analogy, the agroecology-dominant view has agriculture, as a small box, within a larger one labeled agroecology. Two other boxes, one labeled agroforestry, the other labeled forestry, are also inside the larger box. Having three smaller boxes does help subdivide agroecology but at the risk of setting unreal boundaries. Where full land-use integration is the goal, it might be best if the contents of the three internal boxes are figuratively dumped into the larger one. In a more radical view, the smaller boxes, along with their labels, are discarded after being emptied.

6. Cultural agroecology, as a topic of study, was broached by Bradfield (1986), further developed in Wojtkowski (2004).

7. Some tout agroecology as the agricultural solution to global warming or climate change. True as this is, this text is predicated upon the belief that agroecology is superior, economically and environmentally, than conventional monocropping. Agroecology, as a solution to climate change, is only a side benefit.

8. The acceptance of insect loss is not a well-explored topic. One study (Abate *et al.*, 2000) found African farmers tolerate up to 40%. This is the high end of a 0–40% range.

9. Organic gardening, a practice often involving composting, is not to be confused with the organic label. A marketing label defines a class of agricultural outputs, i.e., growth sans manmade chemicals and, because such products can carry a price premium, does encourage chemical-free farming.

10. More on these agroecological subdivisions is found in Gold (1994).

11. Unaccounted for variables running seemingly at cross-purpose do stymie economic studies. This limits the comparative scope (often to agrosystems with the same contained plant species) or, for those brave enough to undertake the more daunting task, the reachable conclusions (as can happen when the systems compared are composed of different species).

12. Agriculture, in this context, includes forestry, i.e., silviculture, and agroforestry. This is implicit throughout this text. Also implicit are treecrops as part of agriculture. Treecrops include bark (e.g., cork and cinnamon), fruits for oils (e.g., oil palms for cooking and industrial oils and/ or bio-diesel fuel) and other uses, saps (e.g., latex for natural rubber, palm wine, and maple sugar), and a host of other non-woody tree products.

13. Although the importance of profit, and in deriving the bottom-line financial picture, is undeniable, this book looks almost exclusively at those economic measures unique to agroecology.

14. For references on the less risky nature of agrobiodiversity, see Endnote 12 of Chapter 2.

15. To name two underutilized, unstudied risk-reducing options, full planting disarray has not been examined, but seems to confer advantage in certain situations (see Wojtkowski, 1998, p. 80). The same holds true with the landscape practice of widely scattering plots to mitigate localized risk (see Wojtkowski, 2004, p. 205).

16. This is departure from the agricultural economics norm when analysis is mostly ex-ante with a clear stop and start between the agronomic data and the economic results. Agroecological economics is far stronger when the analysis parallels or overlaps with agronomic study.

17. Some continue trying to value intangibles, e.g., Pattanayak and Butry (2005), others, including this book, take the perspective that agroecology is a multi-criterion undertaking replete with intangibles.

18. Hydraulic lift was studied by Emerman and Dawson (1996).

19. Although rare, numeric data can be unintuitive. This might occur when theory has not reached, and explained, a practice (see the next endnote).

20. Along these lines, there are exceptions to the rules of biology and ecology. This is demonstrated by the paucity of laws in non-molecular biology and in most versions of ecology. In agroecology, it is not all that unusual to find one-of-a-kind applications, a small percentage of which are based on unique natural dynamics; dynamics are not always fully explained in terms of their agro-complexity and the parameters of use.

2 Lead-Up Agrobiomonics

Proponents of biodiversity publicize the notion that including more plants, productive or otherwise, is a wise course of action. There is much truth in this. As biodiversity is a route to cropping success, a quick journey through a few underlying concepts is good introductory step.

ESSENTIAL RESOURCES

For plant growth, it has been long recognized that certain mineral resources are essential. The three principle elements are nitrogen (N), phosphorus (P), and potassium (K). These, along with light, water, CO_2, and a long list of trace elements, underwrite plant needs. A listing of trace or secondary elements would include iron, zinc, calcium, magnesium, boron, sulfur, copper, manganese, and molybdenum.

Used in various proportions by different plant species, the above constitutes the essential plant resources. With few hard-and-fast rules in ecology, there are always exceptions. One of these, the mushroom, is a plant which can be grown in darkness.

THE LIMITING RESOURCE

Plants do not always benefit from a resource-rich site. Due to a shortage in one limiting resource, the full site growth and yield potential may not be reached.[1] For example, a water-loving species, such as rice, will find water severely limiting if planted in a dry environment.

Across history and geography, countless agriculturists often wrestle with one specific resource shortfall. Those residing in desert climes face a literal do-or-die dilemma in their need to water crops. Less dramatic, but equally troublesome, are declining soil potential due to the exhaustion of a single limiting element.

Where the limiting resources in not overly prominent, this can be a difficult topic, one not always intuitive and quantifiable. Across a growing season, conditions change. What is limiting at one point, may not be at another. When dealing with two or more species, these can share a single limiting resource, as when drought intercedes, or each species will compete for, and find limiting, different resources.

When the means to directly supply the key resource is lacking, circumventing techniques have been developed. Without going into the full range of options, one such solution is a companion facilitative species, one that acquires and makes available a limiting nutrient. This concept, although fundamental in agriculture, is important especially if the limiting resource has a major impact, can be readily identified, and the problem easily amended.

NUTRIENT PROFILES

Each crop species has set resource demands, some want more phosphorus, others require more nitrogen. Whether phosphorus or nitrogen is demanding, rectifying the most demanded resource reveals another resource need.

Another approach is to view essential resources in their totality. This approach ranks the resource needs of each plant species, starting with the resource most demanded, then listing those less needed. What results is a plant essential resource profile.

Once known, the monocrop task is to either (a) find a crop or crop species or varieties where the resource profile of the plant closely fits the prevailing soil resource profile or (b) alter the soil profile to best fit the needs of the upcoming crop.[2] Within a profile context, economic strategies begin to emerge. Finding the best crop or variety puts the economic focus on lowering costs, adding nutrients puts more focus on increasing yields and revenue.

To increase crop yields in the most cost-effective way, a mix of these strategies may prove the best course. This involves finding the crop or variety that best fits residual soil conditions and adding small quantities of nutrients so that the soil profile matches, without overreaching, that of the crop.[3]

There are examples where crops are chosen to fit soil conditions. In early Europe, rye was grown more than the more popular wheat because it better fit the common nutrient profile found with overworked soils. Rotations are a broadening of this strategy. This is where a series of crops, grown across time, are matched seasonally against the ever-changing soil–nutrient profile as residual from the previous crop.

Resource use profiles, crop or soil, can be tangible and quantifiable or intuitive and difficult to ascertain. Where the latter occurs, irrigation and fertilizer recommendations serve as a surrogate for a plant-need resource profile, soil testing provides an approximation of the in-soil resource profile. The problem is more difficult when unlike species are grown together. For this, experience or directed research must answer the soil-multiple species compatibility question.

AGROECOLOGICAL NICHES

It is possible to take yet another step up the complexity ladder. Beyond the essential resource profile, plants operate within ecological niches. These are the environmental requirements of species, crops included. As well as nutrients, soil moisture, rainfall distribution, temperatures, relationships with insects (good or bad), weeds, winds, frosts, and host of other natural forces enter the picture.[4]

Starting with the simplest form, the constrained fundamental niche occurs when like species compete for the same resources. Further along, agro-niches exploit various one-on-one mixed species relationships. From this, two or three species, if well paired, offer numerous avenues for success. This occurs when one species is deep rooted, another shallow rooted and these seek water and nutrients from different sources. Good agro-niche pairings help insure favorable cropping outcomes.

These is yet another step in niche biodynamics. The most complex niche relationships involve growing many plant species in close proximity. In doing so, the niche relationships, i.e., being many and varied, approximate those found in most natural ecosystems.

Whatever the form, understanding and utilizing niche dynamics helps in achieving cropping success. Beyond the monocrop, research and formal guidance is greatly lacking. Much of what can be done is often the providence of the astute farmer, one who has observed the niches requirements of productive species and has tried to accommodate these.

GOVERNANCE

Within a complex operating environment, such as a farm, the question is how to govern the progression from limiting resources on to complex niches. Farmers have the option to completely manipulate the growing environment, as with monocrops, or they can totally surrender control, as when plants are grown in the uncultivated wild.

PHOTO 2.1. An onion patch which, through the constrained fundamental niche of a monoculture, is plant–plant governed.

A high yielding monocultures or well-managed bicultures can require expensive inputs. Numerous intermixed crop species, those raised unmanaged setting, can be a cheap alternative but, in ceding control to nature, these can be difficult to productively manage. At the extremes, there is the option (1) to keep a high degree of jurisdiction (plant-on-plant governance) (Photo 2.1) or (2) let nature have a freer hand (ecosystem governance) (Photo 2.2).

Plant-on-Plant Governance

With lesser levels of biodiversity and careful selection, it is possible secure narrowly defined niche dynamics and direct these toward the economic objective(s). Monocultures are a controlled one-on-one plant

PHOTO 2.2. A garden with potatoes in the foreground which, through density, diversity, and disarray, is ecosystem governed.

setting where each is allocated a share of the available resources. Access to essential resources is regulated by spacing.

A well-designed intercrop still utilizes spacing, but with a reliance upon a few or many favorable niche dynamics. With the classic maize with bean intercrop, these species have unlike soil–nutrient profiles and farmers successfully exploit these differences to obtain high yields.

Problems occur if the niches are not inclusive for all essential resources. With maize and beans, this happens when moisture is limiting and both begin to destructively complete for the one resource.[5] At this point, this intercrop is not economically viable. Other plant-on-plant combinations may have good moisture dynamics but may fall short when competing for other essential resources.

In the above cases, the monoculture or the intercrop is under the direct control of the farmer. Plants are spaced, weeds removed, and other measures are taken to steer the essential resources, in economic sufficiency, to the individual plants.

Ecosystem Governance

As agroecosystems grow in biocomplexity, internal forces take hold. Plants, weeds, and others, grow at will. The landuser, through intent, mains only a cursory involvement. Plant thrive or fail on their own accord or through minimal outside management. Often, the greatest effort is in harvesting. Classic examples of ecosystem governance are pastures containing a mix of perennial grass species or agroforests thick with fruiting trees.

What develops is an agroecosystem where the role of individual species becomes less of a driving force, the agrosystem, in its entirety, incurs the governing role. What happens is that the sum of the ecological parts is greater than what each individual plant and plant species contributes to the whole.

These systems, in exploiting a wide range of niches, produce larger amount of per area biomass. A figure of 238% more than a monocrop has been reported. These systems may also be able to overcome some site limits (as with poor soils or lack of water) and to better protect the site and the contained ecosystem.

In transcending the ecological influence of an individual species, the ecosystem, as an aggregate of multiple small and large effects, can alter the internal soil structure and micro-climate. Among the changes are an increased water-holding capacity and less per-plant evaporation.[6]

Other reasons for the ecosystem dominance involve the contained micro and macro flora and fauna. These blossom in inviting, biodiverse agroecosystems. The flora ranges from microbes to weedy species, each adding a small, a times infinitesimal, contribution to the whole. The same holds for the microfauna, where earthworms and like organisms, both above and belowground contribute, each in their own small way. As each contribution is multiplied by the total, often extensive, populations of these organisms, their all-inclusive influence can be significant. These effects are intuitive, at-times tangible, but not always quantifiable.

ANALYTICAL UNDERPINNINGS

To this point, a few of the broad agroecological concepts are presented. Going down this path requires analysis on which plant-on-plant combination or which agroecosystems are better. Cross-agroecosystem comparison introduces an apples–oranges dilemma, i.e., how to directly assess the output, costs, and/ or risk from interplanting two or more unlike crops.

Basic Measures

In conventional economics, dissimilar outputs are compared by assigning monetary values. In agroecology, comparisons are frequently based on monocultural yields and costs. A powerful concept, this has been expanded into a number of comparative standards.

Land Equivalent Ratio

Central to any discussion of agroecological economics is the land equivalent ratio (LER). Much of the analysis that underwrites agroecology can be expressed through the LER. This is a comparative measure of productivity, but its main strength lies in it being a gauge of efficiency. The LER estimates how efficiently a plant or agroecosystem utilizes the on-site and/ or introduced essential resources.

LER has another strength, intuitiveness.[7] This allows assessments at a glance; the one number providing a clear, unclouded linkage between the biological happenings and yield outcome.

With common underpinnings, i.e., the monocultural outputs for the ecosystems in question, permits the LER to be utilized in cross-agrosystem comparisons. Comparability, i.e., its application as a universal agroecological standard, might be its greatest virtue.[8] Lastly, it is easy to calculate.

The basic equation for an LER, for a biculture is[9]

$$\text{LER} = \frac{Y_{\text{ab}}}{Y_{\text{a}}} + \frac{Y_{\text{ba}}}{Y_{\text{b}}} \tag{2.1}$$

For this, Y_{ab} is the yield of species a grown in conjunction with species b. Y_{ba} is the output of species b grown with species a. Y_{a} and Y_{b} are the yields of monocultures of species a and b grown under like conditions, i.e., soil type, nutrient levels, moisture, climate, etc.

If species a has potential monocultural yields of 6000 kilograms per hectare and, on the same plot, species b, as also in monoculture, yields 4000 kilograms per hectare, these numbers provide the denominators for the above equation (i.e., Y_{a} and Y_{b}). If grown together on the same site and under the same conditions, the two species, a and b, respectively, offer expected yields of 3600 and 2400 kilograms per hectare, respectively, the result,

$$\text{LER} = \frac{3600}{6000} + \frac{2800}{4000} = 1.3$$

Since the LER exceeds one, this indicates a positive essential resource situation. This represents an ecological gain even if the yields of each component species fall short of that obtainable from the comparative monoculture, e.g., $3600 < 6000$ and $2800 < 4000$. A value less than one indicates that the intercrop is not productively efficient, i.e., the co-inhabiting plants are ruinously competitive. For the aforementioned maize–bean combination without any moisture shortfall, LER values about 1.3 are expected.

This equation also works if a yielding species is grown with non-yielding facilitative plants, such as a covercrop. Reworking the above example, one might find, if species a is the main or primary crop and species b is non-yielding and facilitative, a positive association:

$$\text{LER} = \frac{6500}{5000} + 0 = 1.3$$

For this, the presence of non-productive species b boosts the yields of species a by 30% (from 5000 to 6500 kilograms per area).

There are other variations, e.g., as a standard of comparison for monocultures, as when comparing different treatments or management inputs. The LER can also be expressed in triculture form (three inter-cropped species). This version is

$$\text{LER} = \frac{Y_{\text{abc}}}{Y_{\text{a}}} + \frac{Y_{\text{bac}}}{Y_{\text{b}}} + \frac{Y_{\text{cba}}}{Y_{\text{c}}} \qquad (2.2)$$

where a third species c has been added to the mix.

Price-Adjusted LERs

For many uses, selling prices, added to the LER, links in plot happening with market decisions. This can be done by way of the relative value total (RVT). The RVT[10] is computed as

$$\text{RVT} = \frac{p_{\text{a}}Y_{\text{ab}} + p_{\text{b}}Y_{\text{ba}}}{p_{\text{a}}Y_{\text{a}}} \qquad (2.3)$$

For this, Y_{a} is the monocultural yields of species a, Y_{ab} is the output of species a when planted in close proximity with species b, and Y_{ba} is the yield of species b in combination with species a. Additionally, p_{a} and p_{b} are the market values, respectively, for species a and b. The denominator is the same-site monocultural yields multiplied by the value of the primary species. If it is not clear which is the primary species, the one that offers the greatest return from the site in question, i.e., where $p_a Y_a > p_b Y_b, p_a Y_a$ is the denominator.

A calculated RVT would be

$$\text{RVT} = \frac{(\$0.25)(3600) + (\$0.10)(2800)}{(\$0.25)(6000)} = 0.78$$

Despite the use of the same numbers that yielded an LER of 1.3, the outcome with selling prices is decidedly bad ($0.78 < 1.0$). Clearly, given these prices, a monoculture of species a is the better income alternative, i.e. $((\$0.25)(3600) + (\$0.10)(2800)) < ((\$0.25)(6000))$.

Cost Equivalent Ratio

For any economic evaluation, costs are important. In utilizing niches or through governance, there are expensive or cheap options. One must look first at productivity. It also helps of a site, resources and all are being utilized efficiently. This is measured through the LER. Once known, outlays or costs enter the picture.

The absolute costs, expressed in monetary units, are important. Of greater significant is how efficiently the inputs are being utilized. The crude cost equivalent ratio (CER) is a test of this. The core equation is

$$\text{CER} = \frac{C_a}{C_{ab}} \tag{2.4}$$

For this, C_a represents the total costs for a monoculture of crop or species a. This would be the primary, or the most sought after, output. C_{ab} are the total costs for an intercrop of species an interplanted with species b.

The idea is to estimate how efficiently units of input in an intercrop compare to units of input in a monoculture. The C-values are generally expressed monetary units. With subsistence farmers or where labor is the only input, hours worked may be the comparison standard.

If per area monocultural costs (C_a) are \$2 and contrasting polycultural costs (C_{ab}) are \$1, the CER is 2.0. This indicates that, per unit of management inputs, the more complex system gives twice the value per unit of input of than obtainable with a monoculture of the primary crop.

The crude CER is a the stand-alone value for resource-poor farmers, those that have ample land, less in the way of monetary resources, and are seeking low-input solutions. In the hypothetical case presented above, having an agroecosystem that needs one-half the inputs is a good starting point.

RVT-Adjusted CERs

It is more revealing if the CER does not entirely stand alone. The CER may be ungraded in combination with the RVT. The equation for this is

$$\mathrm{CER}_{(\mathrm{RVT})} = \frac{C_{\mathrm{a}}}{C_{\mathrm{ab}}}\,\mathrm{RVT} \tag{2.5}$$

As with the LER, values for the RVT-adjusted CER that are greater than one indicate polycultural superiority. Values less than one show that a monoculture is more efficient with the same inputs and outputs.[11]

Take the case where a system has a CER of 2.0 and an RVT of 0.78, the CER, RVT adjusted, will be 1.56. The breakdown denotes RVT that is not promising (as compared with the crop monoculture), but when management inputs are considered with the LER, the system still shows significant gains. In this case, the $\mathrm{CER}_{(\mathrm{RVT})}$ value (1.56) shows that the cost gains be of greater worth that the income losses.

This version of the CER is a stand-alone value. These can be employed to cross-compare, and make generic judgments, on some very diverse agroecosystems. As with the LER, intuitiveness and cross-agroecosystem comparability make these a universal standard.[12]

Economic Orientation

Agroecosystems may be less of interest for their outputs, more for the fact that the outputs, even at a reduced yield levels, can be produced with some degree of input or cost efficiency. As such, the economic orientation ratio (EOR) is a further refinement of the LER and the CER.

High levels of output are nice but, given diminishing marginal gains, the last unit produced can be expensive (using a per unit valuation). Instead of seeking more, some farmers, especially those without money for inputs, may seek to produce at least cost.

The EOR can be determined through the following equations:

$$\mathrm{EOR} = \frac{p_{\mathrm{a}}Y_{\mathrm{ab}} + p_{\mathrm{b}}Y_{\mathrm{ba}}}{p_{\mathrm{a}}Y_{\mathrm{a}}} - \frac{C_{\mathrm{a}}}{C_{\mathrm{ab}}} \tag{2.6}$$

or

$$\mathrm{EOR} = \mathrm{RVT} - \mathrm{CER} \tag{2.7}$$

As stated with the RVT and CER, Y_{a}, and C_{a} are the yields and costs of species a monoculture, while the Y_{ab}, Y_{ba}, and C_{ab} are those resulting from an intercrop of species a and b. Note that, for the CER, the unmodified version is employed.

If the RVT is greater than the CER (i.e., with a positive EOR), a system is revenue oriented. If the CER is greater than the RVT (i.e., with a negative EOR), a system is cost oriented. Ideally both will occur and reductions in costs will lead to increased yields. In this case, the system will return a zero or near zero value. Being the rare case, this is of less immediate concern.

In example, an agrosystem with an RVT of 0.78 and a CER of 2.0 shows an EOR of −1.22. Being negative, this system is clearly cost oriented.[13]

The equation produces two useful bits of information: (a) the range and (b) whether negative or positive. The sign, positive or negative, indicates economic orientation. The spread range is an indirect indication of profitability (i.e., revenue minus cost).

In the vast majority of cases, there are two separate economic strategies. The first is *revenue orientation*, adding inputs as long as the cost of the added input is less than the value of the outputs gained. The second is *cost orientation*, reducing inputs as long as the value of the reduction is less than the cost savings being realized.

Revenue orientation, achieving the highest LER, can require a productive secondary species. The increased revenue comes from the primary species plus sales of the secondary plants. High-input, high-output monocultures also qualify. Extreme revenue orientation finds favor where markets are strong and agricultural land is in short supply. If market value for the primary species is weak, the difference may be made up by activity focusing on and marketing the secondary outputs.

With cost orientation, the secondary species tend to be facilitative, offering little in the way of secondary outputs and mainly dedicated to reducing costs. There are low-input monocultures that, through input efficiency or input replacement, qualify as being cost oriented. In either case, facilitative biodiversity or a low-input monoculture, yields are not expected to be high. These find favor when farms are large and do not have sufficient resources, e.g., costly inputs and labor, to full satisfy all plots. Cost orientation is also favored when selling prices are low and the low cost of production can still bring on system profitability.

Orientation is a powerful concept in agroecology. In illustration, proposing highly revenue-oriented systems to land-rich, resource-poor agriculturalists can be a recipe for failure.

Figure 2.1 shows a dual orientation scenario with two cropping possibilities. For each, the crop is the same but, instead of outside inputs (right curves), a second agrosystem replaces the inputs with low-cost, natural controls (left curves). The classic examples are, for

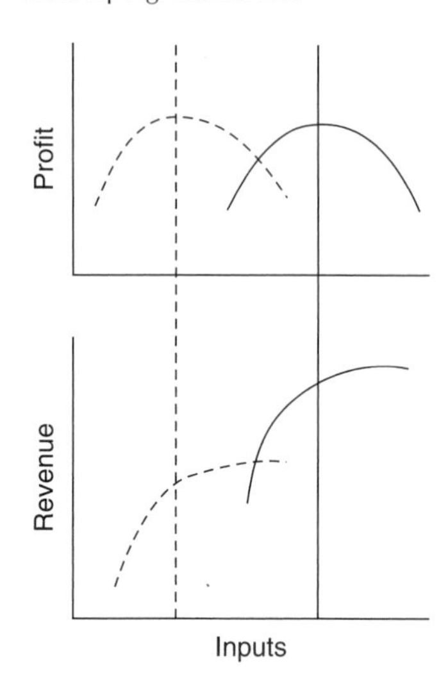

FIGURE 2.1. Revenue (right) and cost (left-dotted lines) orientation with two distinct systems. Each of these produces the same profit (top), but at different levels and with different amounts of inputs (bottom).

the right-hand curve, resource-demanding monocultural coffee and, for the left-side curve, low-need coffee grown beneath trees. The economic question is which is more profitable and, equally important, which better fits the economic and environmental constraints of a farm.

Risk

Risk aversion is also part of the biophysiology of agroecology. Whether a cash-based or a subsistence farm, landusers seek to eliminate or minimize the risk of crop failure. Systems based on biodiversity are, in general terms, better safeguarded against failure or severe loss than conventional monocropping.[14]

The standard for evaluating risk is often subjective, still some comparative measure is of value. Risk has two components: (1) the frequency of anti-output events and (2) the severity of each event. Using these, the basic equation to determine the threat assessment (TA) is

$$\mathrm{TA} = (1 - H^r)^L \tag{2.8}$$

There are three components of this equation:

1. The frequency of a counter-crop happening (H). This is the interval at which each event occurs, e.g., once each 5 years (0.20), once each 10 years (0.10), once each 25 years (0.04), etc.

2. The severity of each event (L) measured as percent (%) of the crop lost to each event, calculated using a decimal, e.g., 20% of the crop lost (0.20), 50% lost (0.50), etc.
3. A risk-expectation factor (r-factor) takes anticipation into account. The idea behind the r-value is that events that occur with some frequency, e.g., every other year, once every 3 years, etc., are anticipated. Being foreseen, measures are usually taken to lessen their severity (e.g., the most susceptible crops or varieties not planted, money set aside, and/ or alternative, life saving crops are in place on other plots). The r-value, in the 0–1.0 range, is set high if frequent events are somehow unanticipated, low where these are expected. This is the human element in risk; risk seekers merit a high value; risk avoiders, a low value.[15]

An simple example, where a negative event can be expected, on average, every 10 years, this is expected to destroy 50% of the crop. These numbers give a TA of 0.83. If, at the same interval, 100% of the crop is destroyed, the measured outcome is 0.69.

The TA measure lacks intuitiveness, a basic requirement in agroecological economics. The risk index (RI), as below, corrects this:

$$\mathrm{RI} = (1 - \mathrm{TA}) \tag{2.9}$$

Clearly, no risk is zero risk. This is true with the RI. The range goes from zero risk to the risk extreme (1.0) where all the crop is lost every season. Illustrating the above calculation, destruction of 50% of the crop each 10 years gives a RI of 0.17 whereas a 100% loss each 10 years is more severe, with an RI of 0.31. A 100% crop loss every year is a 100% risk (1.0 on this scale). This scale can be adjusted, as is done in latter chapters, for specific purpose.[16]

ENDNOTES

1. There are a number of theories that underwrite a single limiting resource *via-a-vis* the other essential resources. Without going into detail here, there are discussions in ensuing chapters along with visual articulation (Figure 10.1).

2. Precision agriculture seeks to fortify the soils in any one micro-location *vis-a-vis* crop needs by taking into consideration of the light and water availability. As a cost containment measure, the idea being not to waste money by over applying any one nutrient.

3. Overreaching in resources application is a problem involving marginal gains (see Chapter 10, section 'Marginal Gains').

4. The concept of the niche does, at times, vary from that proposed here. Although a complete discussion is outside the realm of this text, more can be found in Liebold (1995).

5. Competitive problems within the maize–bean intercrop appear when seasonal rainfall drops below 400 millimetres (Rao, 1986).

6. With many reporting, Tilman *et al.* (2006) provide the 238% figure while confirming the ability of such system to overcome site limitations. This work was for a biomass harvest. When fruit and other treecrops are the output, questions remain on the overall level of outputs. Despite measurement difficulties, it is generally assumed to be good or very good.

7. Keeping in mind that agroecology is and will continue as more of an art than a science, intuitiveness comes into play in deciding which design variables provide the greatest worth and should be looked at first. For example, does a proper plant spacing increase the LER by 1.0 or by 0.1? Answering such questions, in an instinctive way, helps in improving local cropping systems.

8. As an agroecological measure, it is critical that LER be intuitive and offers universal cross-agroecosystem comparability. Except where clearly noted, variations should not be endorsed that lose these characteristics. The LER is a strong concept, enough so that there may be potential in developing LER-based comparative statistics. Work along these lines could resolve the comparative dilemmas often found in the literature (see Endnote 7). Therefore, it would not be a step too far to suggest that all multiple cropping research be presented in LER units.

9. The LER was first proposed by Mead and Willey (1980).

10. The RVT comes from Schultz *et al.* (1982).

11. It would be equally proper to refer to the RVT-adjusted CER as the CER-adjusted RVT ($RVT_{(CER)}$).

12. One should be careful in formulating far-reaching versions of LER. If too much is included, intuitiveness and the LER as a gauge of efficiency will be lost. This happens when the LER, CER, and the like are coupled with net present value (for more on the temporal dimension, see Chapter 6).

13. Taken together, the progression of numeric examples in this chapter paint an intuitive and insightful economic picture. The two co-planted plant species, when intercropped, are site and resource compatible (the LER = 1.3), but the relatively large difference in the output selling prices does not economically inspire (with an RVT of 0.78). What makes this system fully viable and of economic interest are the cost savings (the $CER_{(RVT)} = 1.36$). The EOR (at −1.22) suggests a system best promoted in rural, land-plentiful landscapes where labor and like inputs are scarce.

14. Among those reaching the conclusion that agrodiversity is less risky than a monoculture are Lotter *et al.* (2003) and Dapaab *et al.* (2003). Risk is not always a function of one plot or one crop, risk is often distributed across a landscape function (see Chapter 15, section 'Spatial Concerns').

15. The *r*-factor also gauges the added impact of prolonged, negative events, such as cross-seasonal droughts.

16. The data for scaling risk maybe a lot closer than the literature suggests. Two possibilities exist: (1) analyzing yields by means of a general crop rainfall (or temperature) yield response function and regional weather data (a sample regional yield response function is found in Glover, 1957) or (2) scaling through crop failure insurance payouts.

3 Vector Theory

Conventional monocropping is predicated on certain beliefs. First among many falsehoods is that plots of a single species afford the best economic outcome. The supposed superiority of the one-crop solution is true only because over a century of research has refined this agroecosystem type to high degree of sophistication. The falsehood is exposed in that, after all this work, a satisfactory conclusion has not been reached; environmental problems still come to the fore, crop failures still persist, and harmful insects and infecting plant diseases have yet to be conquered.

This is not to say that monocropping is to be shunned. Quite the contrary, it is possible to build upon this research, but in other directions. There are many avenues from which to choose.

VECTORS

In dealing with complex ecosystems, a single analytical equation cannot often express what is happening. Alternatively, the explanation may depend on or be derived from sets or categories of agroecological mechanisms or treatments. The sets come together to invoke, and explain, a complex outcome. Simple in concept, these offer, by varying the strength and dynamics of each set, individualized interpretations.[1]

This broad analysis, that of mechanistic subdivisions, holds in agroecology. These are the agroecological vectors. The purpose is to subdivide agroecology into understandable sets. The resulting vectors categorize the agrobionomic (or agrobiodynamic) mechanisms.

The Base

In unraveling the terminology, it is easier to open with an unmitigated monoculture, one without agroecological additions. This fundamental or base system is composed of a single, genetically pure (clonal), plant species without rotational planning, without nearby ecologically bolstering ecosystems, unsupported in any other ecological or non-ecological manner. There are no inputs, labor or otherwise, except in planting and harvesting.

The common assumption is that, without inputs, this system is unsustainable and naked to threats, climatic or otherwise. Therefore it must be manually supported and/ or ecologically extended to insure sustainability and additionally, it must be hardened against the various risk factors.

Agroecological Vectors

Agroecology offers a number of ecological vectors or solution directions which can increase production, add sustainability, and/ or reduce risk for the threat-naked clonal monoculture. Each vector has a direction and force. Each vector alone is capable of making a large contribution converting the unadulterated, unsustainable monoculture into something field usable. All applications depend on vector combinations.[2]

The vectors are:

- genetic improvement
- varietal
- microbial
- agrobiodiversity
- biodiversity (facilitative associations)
- rotational
- cross-plot
- location
- physical land modifications
- ex-farm inputs
- environmental setting.

Genetic Improvement

Genetic improvement is a facet of agriculture. Most agricultural plants have and continue to undergo a domestication process. This begins when plants are transformed from a wild state to become field-ready agricultural additions.

One case is particularly telling. When observed in the wild, the untamed parent of maize, teosinte, is not recognizable as the most common of field crops. Improvement across many millennium has changed this.

All agricultural plants undergo a similar experience, some being closer to their natural form, others have gone through considerable domestication-related change. The goal may be higher yields, an expanded growing range, better nutritional content, improved taste, ease of propagation, less difficult harvests, or less in-field risk.

The yields of wild parents seldom approach their domestic offspring. Some, notably wild lentils, are not productive enough for a formal planting.[3] In another gain, the domestic versions do not shed their seeds, as a result, fewer kernels are lost and harvests are expedited.

Domestication is a slow, often erratic process if undertaken by farmers, speeded up when research driven. An example of a concerted domestication effort was the green revolution. This achieved considerable success by making common crops better suited to conventional farming practice.

The genetic vector may have underwrote the green revolution, but this was employed in tandem with an ex-farm inputs vector, i.e., success came, not because the improved plants could survive totally on their own genetic merits, but because the new varieties relied on liberal doses of farm chemicals (insecticides, herbicides, fertilizers, etc.).[4]

Furtherances in genetic modification have accelerated this development vector, armoring cultivated plants against adversities, e.g., climatic variation, disease, and insect attack. There are disadvantages. Often cited is the all-eggs-in-one-basket danger where clinks in the genetic armor many be exploited by a range of injurious organisms. The end result, for each of the commercial crop species, may be one or a few 'super' varieties.

Varietal

The domestication process involves selection, initially from a wild state to useful agricultural plant. Along the way, the plant often branches into numerous varieties. Multiple varieties are an exploitable agroecological resource.

Wheat had been shown to have more than 600 varieties.[5] For other staple crops, e.g., potatoes, maize, and rice, similar tallies are probable. Some of this occurs when a variety is moved to a new location, where the plant may eventually acclimatize. In the process, another variety is born. Maize, originally a tropical plant, now offers short season and frost resistant types, allowing growth in cooler regions.

Varieties of common fruits and vegetables, e.g., tomatoes, beans, apples, and pears, are a quality-of-life resource, furnishing the human diet with flavor, nutrients, and color. A case in point, common vegetables (e.g., carrots and potatoes) and other crops (e.g., cotton) come in different hues. This is a variety-related characteristic that should be more utilized as, for the edible crops, color lends taste and improves the nutrients content and, for fibers, this can circumvent the dyeing process.

An example where varieties manifest considerable outward variation is *Brassica oleracea*. This one plant species offers farmers and consumers cabbage, broccoli, brussel sprouts, cauliflower, kohlrabi, and collards. Each variety is an accentuation of different parts of the same plant species, e.g., cabbage emphases the leaves, caulifower the flower, kohlrabi the stem, etc.[6] This is not the only use of this vector.

New species continue to evolve and, with agroecological trepidation, others are lost. As a development vector, it is desirable to select, often among many, the absolute correct variety for the market, climate, soil profile, and to resist nearby threats.

Also part of this is the mixing of varieties with complementary characteristics. A drought-resistant variety may be interspersed with water-demanding type or insect resistant types can be intercropped. Rice yields have been dramatically improved (in the range of 40%) by mixing varieties, thereby thwarting herbivore insects.[7]

As a vector, this is little utilized. Studies are lacking and there is almost no guidance on practical application, i.e., how to classify and utilize this considerable resource to address agroecological problems.

Microbial

Micro-fauna, in the form of plant diseases, is potential danger to agriculture. However, microbes are large group of organisms. With so much diversity, many serve to advantage.

Well recognized are species associated, nitrogen-fixing mycorrhizae. Pine trees on poor soils require a mycorrhizae association for growth success. This is a one-to-one, microbial-species interrelationship. Prominent with pines, this type of one-on-one relationship, microbes helping plants, exists less noticed with many species. Unassociated microbes, those not found in conjunction with a single plant species, complete the picture.

In-soil microbes perform many ecological tasks. As well as fixing nitrogen, some make other nutrients available. This can include breaking down chemical compounds allowing the elemental nutrients to float free or breaking down rocks to release those nutrients physically immured. Microbes also help ecosystems and plants hold water.

It is possible to inoculate plants against adversities, most predictably against herbivore insects and plant diseases. This works by inflecting and liquidating insects and other plant attacking organisms, diseases included.

A class of organisms, endophytes (in-plant living fungi) safeguard in other ways. There is temperate protection where plants endure higher temperatures than otherwise possible. Other endophytes help plants withstand drought and may even reduce the sunlight requirement.

This is a mostly unexplored aspect of agriculture. The potential is mostly unrealized, the microbial vector proposes utilizing the vast array of microbes (bacteria, fungi, endophytes, etc.) as specialized and general-purpose agroecological tools.

Agrobiodiversity

The raising of multiple agricultural species, in close proximity and for mutual benefit, is an important part of agroecology. For some, intercropping defines agroecology. Also under this agrobiodiversity (or agrodiversity) heading comes various forms of agroforestry and, to a far lesser extent, the multi-species silvicultural plantations. The key requirement of this category is that all intended species provide an economically interesting output.

In putting forth the strategies behind the agrobiodiversity vector, a number of subcategories come to the fore:

(a) *Archetype agrodiversity*: This is where every included plant species is integral in the planned economic outcome. The common case is intercropping where maize, bean, and squash provide mutual benefit and three harvestable crops.

(b) *Expanded agrodiversity*: As diverse agroecosystems are established, it is possible to add plants without altering the ecological character or the economic intent of the overall system. The common example is when many forage species coexist within a pasture.

(c) *Agro-enrichment*: Once an agroecosystem is established, light, water, and soil nutrients may be under-utilized. When this occurs, it is possible to insert a short-duration species without invoking a strong and negative competitive influence against the primary crop(s). This mechanism is imposed when early maturing radish or lettuce are planted between recently established tomatoes (Photo 3.1).

(d) *Casual agrodiversity*: It is not unusual for useful a plant to naturally occur. Squash, from a previous planting, may sprout amidst maize. This unplanned entry is allowed to remain if it does not interfere with the primary species. The same may happen when, upon clearing land for a planting, a few useful trees or shrubs are found. Rather than removal, the plants may be incorporated, if few in number, into the new agroecosystem as a casual addition.

(e) *Supplementary agrobiodiversity*: In nature, unused niches are soon filled, the same should be true with agroecosystems. This holds with casual agrodiversity but applies equally well to inserted plants. In a forest, vines occupy a niche and, in a mimic of nature, commercially marketable rattan vines can be grow over the top of an close-spaced orchard, treecrop, or forest-tree plantation. Another unoccupied niche is actuated when truffles are grown beneath forest-tree

PHOTO 3.1. Lettuce with tomato where the lettuce will be harvested early in the season allowing the tomato full access to all essential resources.

plantations. In contrast to agro-enrichment, there many be no clear resource opening (i.e., unused water, light, and nutrient resources), but still exists an opportunity to squeeze in another plant species.

Biodiversity

Rather than employing a mix of agricultural species (agrobiodiversity), non-productive plants can assist agricultural production through favorable interspecies mixes. There are advantages in this; it can be hard to find a plant that is non-competitive for light, water, and in soil nutrients while producing something of market worth. By seeking a non-agricultural species, the choice is wider and the possibility of finding a good pairing, one that does not overly detract from yields, high.

As with agrobiodiversity, the subdivisions are:

(a) *Facilitative biodiversity*: Purely facilitative plants can discourage herbivore insects, slow the spread of a plant disease, smother weeds, improve upon the water and nutrient gathering ability of the primary species. Gardening books list many such species, among which are decorative plants. An example, temperate gardens may accommodate marigolds (pretty flowers and nematode control).

(b) *Expanded facilitative biodiversity*: Rather than achieving one or two facultatively simple tasks, it is possible to expand upon the number of facilitative species. These species are expected to ecologically unite to counter a range of agricultural problems. The notion here is to use one species for insects, another for erosion control, etc.

(c) *Casual biodiversity*: Without being invited, plants often colonize a site. Many can play a facilitative role provided they are compatible with the primary species (one or more). As with casual agrobiodiversity, these are allowed to stay if their perceived value exceeds any losses incurred.

Rotational

Soil nutrients, insect populations, and other productive factors are affected by sequencing crops. This approach puts more of the ecological emphasis on cross-seasonal, cross-species dynamics. Rotational gains are time tested which often, but not exclusively, involves crop sequences where the soil characteristics after harvest best meet the nutrient requirements of the upcoming crop. Outside of the nutrient gains, rotations are employed to disrupt the life cycle of crop-eating insects, crop-killing diseases, and some weed species.

Cross-Plot

Auxiliary systems, in the form of windbreaks, are found in many landscapes. These prevent wind erosion, keep plots from drying, and protect plants from the thrashing and stem rubbing effects, all of which reduce overall yields. Besides being prominent in many landscape, these typify cross-plot associations.

Although windbreaks may be the most recognized auxiliary system, other nearby ecosystems can realize similar results, that of countering otherwise negative influences. In another example, movement corridors, strips of natural vegetation between plots, allow predator insects, those that eat the herbivore types, to quickly populate farm fields.

Instead of relying on ecologically self-contained plots, a range of objectives can be addressed through interplot associations. One-on-one cross effects are the basis of the cross-plot or one-on-one landscape vector.

Location

Instead of a varietal matching to make crops compatible with a site (a crop-first approach), crops can be placed where they grow best (a plot or site-first approach). Topography, soils, micro-climate, etc., are utilized in this quest, i.e., to reconcile the crop with the location.

The idea being that, rather than force the issue, plant something that naturally thrives in the site as presented. Soil nutrients, moisture content, ambient temperatures, insect pests, length of the growing season, and sunlight intensity are among the premeditating factors.

The difference between the site-first (locational) and crop-first (varietal) approach is vague, but real. Take a wheat farm where one plot contains poor, sandy soil; with a crop-first, varietal approach, a best wheat variety would be found to accommodate the soil deficiencies. Alternatively, a poor-soil tolerant plant, such as rye, would be planted. Finding the best variety is a plot-first approach, the latter, situating the best crop, is locational.

There are a number of applications. This can be micro-site, e.g., one crop on the top of a mound, another on the side, yet another in the small, between-mound valley. This approach also exploits larger topographical features. The elevational gradients on hillsides often figure prominently in a locational planting.

Physical Land Change

In some areas, especially where climate is an obstacle to crop growth, physical changes in land shape can stretch the limits nature imposes. Examples are not hard to find, terraces allow sustainable yields on steep hills or mountainsides, paddies permit water-loving crops, such as rice and cranberries, to thrive on an otherwise dry site. Physical land modifications, those where earth is moved, can go far in promoting agricultural goals.

Ex-farm Inputs

Commercial agriculture, as practiced by western societies, has come to rely on ex-farm inputs. These can range from the environmentally benign to those that are environmentally toxic.

There is a whole class of chemicals, often referred to as home remedies, that have been long utilized to promote crop yields. Most often these are completely benign or the negative consequences are of short duration.[8]

On the other side are the ex-nature, synthetic compounds that, unless handled well, can cause problems. These include man-made insecticides, fungicides, herbicides, and fertilizers. For many, this is an enticing line of development.

As already mentioned, the ex-farm approach has been fused or allied to forward farm goals, first through the green revolution, later through genetic engineering. Development continues, e.g., where a genetically developed, herbicide-resistant crop is weed managed through weed-destroying chemicals. The notion is to reduce herbicide amounts and costs by targeting susceptible weeds growing amongst the herbicide-resistant crops. On the negative side, the weeds, also genetically evolving, begin to resist the herbicide, requiring ever greater amounts to achieve the desired control levels. Another observed side effect has the herbicide-resistant crop emerging as a weed threat.

Environmental Setting

As a catch-all category, with many diverse influences falling under this heading. The basic notion is that, through changes in everyday agricultural practices (i.e., the setting), the internal natural dynamics, economics, and risk factors of agroecosystems can be improved.

The underlying mechanisms can be broad, not targeted, indirect, and not often well understood. Despite an intangibility and a calculation difficulty, the setting can be economically significant.

In one case, the application of animal manure controlled aphid populations. Chemical fertilizers did not achieve the same.[9] The influence here might lie in the triggering of, or establishing a mini or subecosystem, on that results in a blossoming of microorganisms. These, in turn, supports larger organisms. Taken together, this bloom in useful microorganisms benefits crops. Similarly, compost, if well applied, can instigate a crop-favorable environmental setting.

This is not exclusively with microbes. In a well-documented case from China, the citrus ant is a keystone species that establishes a control matrix or control hierarchy where, not one, but all the citrus-damaging insects are regulated. Management under an environmental setting heading keeps this matrix on course.

Insect-on-insect control is only one part of the picture. Birds, either domestic fowl or wild species, can be encouraged to pick off insects, reducing or eliminating an insect problem. For this, bird-friendly surroundings are part of the environmental setting.

There are other forms of environmental setting established through management. Where accompanying grasslands are a haven for predator insects, the type of mowing does change the setting and outcome. High-speed rotary mowers pulverize both good and bad insects, reducing the effectiveness of the grass strips.

The plowing method can also establish an environmental setting. Rotary plows, those that grind up soil-improving earthworms, are counterproductive in any agroecological context. Less intrusive moldboard or disk plowing, or better yet, no-till techniques, represent an amelioration on an otherwise destructive force.

Allying Vectors

In combination, these provide the tools (through subcategorized agrobionomic mechanisms) to enhance and protect various crops on varying sites, climates, and against a range of threats. Among the specific tasks required are high yields into the distant future. All kinds of threats, harmful insects, plant diseases, high winds temperature extremes, etc., must be overcome. No one vector does this. Most of what occurs in agroecology involves the allying of vectors.

An example is wind damage. Genetic or varietal approach can address some aspects; stems can be made shorter and stronger to resist lodging, the epidermal layer can be thicker to resist interstem rubbing, and plants made drought resistant to resist drying. Or a windbreak can be installed around the affected plot. In this case, a single auxiliary system, the windbreak, may be substantially easier than attempting to enhance the internal plant characteristics or finding a wind-resisting variety. In allying, both approaches are utilized.

There is the previous described green revolution and genetic engineering. The net effect was to make crops more responsive to, and reliant upon, ex-farm chemicals. Work could have proceeded in other directions, e.g., making crops more responsive to naturally occurring, in-soil, mineral resources or better for inclusion in an intercrop.

DESIGN

Starting with the base monoculture, it is possible, through vectors, to proceed in any number of directions, i.e., to custom design an agroecosystem. In doing this, specific objectives (i.e., yields, profits, environmental intent, etc.) can be sought, the economic emphasis (i.e., revenue, costs, and/ or risk) can be directed and subsequently managed.

DESIGN VARIABLES

Within agroecosystem design, there is a lot to deal with. In this chapter, the usefulness of vectors is explained. These are at the top of the hierarchical layering. Being at the top, these function more as abstract influences.

At the bottom are the design variables. These make things happen. A listing would include:

- Plant species
  - Primary species (generally one)
  - Secondary species (one or more)
- Spatial patterns
  - Pattern arrangements
  - Dimensions
  - Canopy patterns
- Temporal adjustments (intraseasonal)
- Management options/ inputs
  - Labor inputs
    - Tillage
    - Establishment (planting)

Plant maintenance (with shrubs and trees)
Weeding
Supplemental inputs
Fertilizers
Insecticides
Herbicides
Fungicides
Irrigation

DESIRABLE AGROECOSYSTEM PROPERTIES

The combination of vectors coupled with design variables, not forgetting the limiting resource, nutrient profile or niche approaches and the form of governance (in the prior chapter) compile into a plenitude of agroecosystem designs. Whatever type of agrosystem is installed, it is bound to have strong and weak points, e.g., well harden against some threats, less soundly braced against others. It is this mix of properties, strong and weak, that ecologically and economically define an agroecological technology or agrotechnology. The strengths are the desirable agroecosystem properties (DAPs). These include:

- yields (short and long term);
- controlling weeds;
- resisting soil erosion;
- combating plant-eating insects;
- withstanding plant diseases;
- deterring destructive breezes or high winds;
- repelling ruinous fauna (birds, snails, mice, etc. as well as larger grazers);
- mitigating droughts and inundations;
- countering temperate extremes;
- accomplishing a host of other smaller and less noticed tasks.

DAPs are a formalized statement of the pros and cons of any proposed system. These help in a number of ways. Adoption is easier if the DAPs of a system match the site and economics needs the user requires. Take the case of hillside plot, erosion resistance is DAP and, lacking this, a proposed agroecosystem would not be a strong candidate for adoption.

Figure 3.1 shows how the DAP concept is employed. Above has comparative measure of desirable properties for three distinct agroecosystems (designs a, b, and c). The one selected best matches what the user wants.

The goal of any change, minor or dramatic, is to keep the strengths and fortify the weaknesses. Taken to its fullest, DAPs carry agroecology

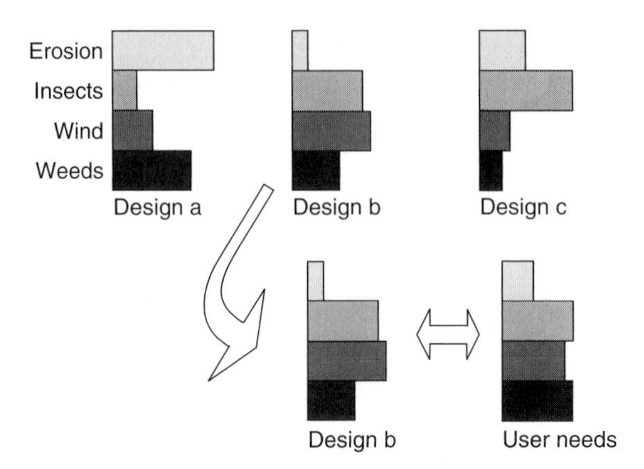

FIGURE 3.1. The application of DAPs where each agroecosystem design (a, b, and c) combats threats to varying degrees. The one selected (in this case design b) most closely matches the needs of the landuser.

to a higher plane. Part of this is practical, part is economic. There is the need for plot optimization, i.e., taking something useful and deriving something better. Agroecosystems, as defined through their DAPs, shortcut a lot of agroecological complexity, allowing users to single out those systems that are initially strong and have the capacity for improvement.

AGROTECHNOLOGIES

To extract the most from the vast array of possibilities, one must find a unique combination of vectors and variables expressed through the DAPs. If well chosen, those are close to, or on the right track for, optimized on-farm solutions. The agrotechnologies are an intermediate step between the vectors and the variables.

These represent composite of vectors defined, ordered, and prepackaged so that farmers can take these as starting points for their own use. Most of the existing agrotechnologies are an offshoot of field practice where local farmers found a better way to apply ecological principles or, alternatively, researchers have done likewise, learning how to exploit agrobiodynamic mechanisms.

At times, the line that separates the design variables from the agrotechnologies is somewhat fuzzy. Clearly species, primary or secondary, and spatial patterns are elements in designing agrotechnologies. Inputs, such as synthetic fertilizers, are one dimensional, i.e., not expected to do much more than boost yields. As such, these are design variables. Green and animal manures may be expected to do more, e.g., the already mentioned case where manures control aphids. As such, these may cross the line to become a low-level, add-on agrotechnology. Another category, discussed in this chapter, is the agroecological add-ons.

Principal Mode

The most important agroecosystems are principal mode. These provide the farm outputs and are the economic driving engine for all farm landscapes. Plots of any economically useful crop, mixed or otherwise, are principal-mode agrotechnologies.

Principal-mode systems can be viewed as a multi-dimensional continuum of agroecosystems, each a slight species (one or more) or spatial variation of another. For example, a plot may contain hundreds of plants of species a, one for species b, or the reverse may occur. Within this continuum dimension, there can be hundreds of plants of both species, one plant of each, any combination in between. Other continuum dimensions have two or more species in various spatial patterns, e.g., row, strip, cluster, boundary, etc.

The resulting agrosystems can be fully productive (agrobiodiverse) or partially productive (biodiverse), i.e., both species a and b can be productive or one can be facilitative, the other productive. Some combinations of species, spacing, and spatial patterns (e.g., see Figures 4.2 and 5.4) are more potent than others in providing outputs; in terms of their essential resource efficiency as captured through unique biology, ecology, agrology, etc.

Figure 3.2 conceptionally diagrams this. Each of the lines that crisscross the species a and b density domain represent variations of one spatial pattern. This can be a single plant of one species (usually a tree) amongst many smaller species or many plants of equal size.

Of interest are the circled portions of some lines (Figure 3.2). These are the planting ratios that offer a high degree of essential resource-use efficiency. It is these designs, with their specific species (two or more), density ranges, and an effect-capturing spatial pattern, that, if economically intriguing, merit the designation of a core or base agrotechnology.

Not all species combinations and spatial patterns have, with their density reach, points of economic interest (e.g., lower line, Figure 3.2). These would not be considered agrotechnologies.

Another dimension for this classification system are the temporal agrotechnologies, i.e., adding a time span to species, spacing, and spatial pattern. As expected, there are different across-time patterns that confer productive advantage. If exploited to full utility, these can boost productivity, reduce costs, all the while mitigating some risk factors in part or all of a cropping sequence. Some sequence formulations work

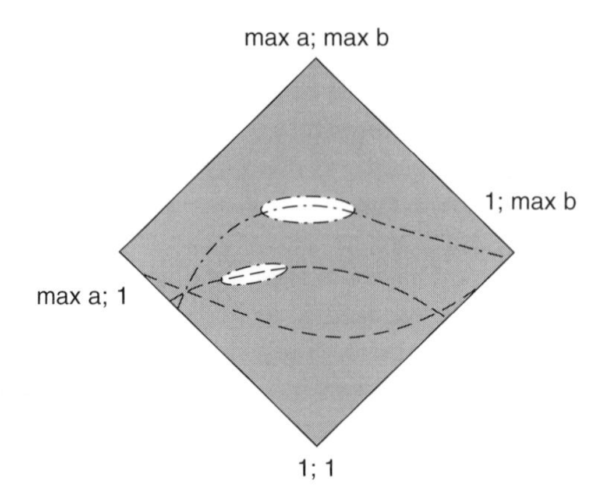

FIGURE 3.2. An overview of the two-species density domain where there is, at the minimum, one of each species (1; 1) and some unknown maximum (max a; max b). The crossing lines delineate various spatial patterns. The circled portions represent points with high resource-use efficiency that underwrite most, but not all, agrotechnologies.

well, others don't. Again, those that confer proven advantage merit an agrotechnological designation.

Besides being the economic driving engine of a farm landscape, the individual agrotechnologies are decision points. It is important that the mix of DAPs match those required on-site. Although important, factors other than the DAPs also come into play. Other agroecosystems can and do, through DAPs, sway the direction taken.

If a farmer has pastures, there is little incentive to adopt a cropping system where animal forage in secondary output. The DAPs of the proposed system must be there, e.g., feed for hungry cows or goats. Once this match is made, there must be a clear and easy to path from what is (e.g., pastures) to what could be (e.g., pastures plus additional forage). It can be the path, more than the DAPs, that dictates which agrosystems eventually find use.[10]

Adoption paths aside, having predetermined agrotechnologies from which to plan does shortcut the decision process. There is the yes-or-no regarding each agrotechnology. For many the choice, to use or not, can, and often has been, distilled to a few key decision variables. An example is revenue or cost orientation. Users desire one or the other and choose agrotechnologies accordingly. It helps greatly that the agrotechnologies have been presorted along economic orientation lines.

In many cases, the standard agrotechnology design often needs refinement. Byway of agrotechnologies, it is easier to proceed along already explored tracks than to embark unguided. As an economic question, the base requirements, user needs and site, for each agrotechnologies are paramount, and the first step, in selecting an appropriate system.

Auxiliary

The concept of the agrotechnology can be expanded. This can include not just the design of cropping agrosystems, but to any in-field practice that confers economic and/ or ecological well-being within an agricultural setting.

Auxiliary ecosystems do not have agricultural outputs as the main goal, instead these provide productive, ecological, and environmental support to nearby principal-mode systems. Many auxiliary designs offer no salable outputs although these do have DAPs.

The classic example, the windbreak has the primary task of protecting neighboring crops from wind-related damage. These systems, with the goal of improving nearby plot economics are best implemented under the cross-plot or landscape vector heading.

Add-Ons

There are a class of add-ons that further agroecological formulation. These can be major change, such as incorporating rotations or, slightly lesser, such as interchanging the plant variety. Some seemingly minor changes can produce major results.

Critical in defining an agrotechnology is the partnership with other variables which, through an integrating design, are intended to produce an auspicious, optimized or near-optimized, outcome. A slew of seeming minor changes or add-ons, in set situations, can be a highly favorable agroecological additions. Agricultural birdhouses, coupled with bird-friendly surroundings and a plan of attack, are a mechanism to attract and keep insect-eating birds. As such, birdhouses are an add-on agrotechnology (Photo 3.2).

Advantage is also gained through actual land modifications. This often involves some digging, the object is a field or plot in better physical form to support crops, a sure ecology, and the output levels envisioned. The modifications that fall under this heading include terraces, infiltration ditches, and paddies.

Other variables include plant spacing. Because these are minor inputs, few cross the line to become agrotechnologies. These still remain as highly useful add-ons. These also fall under an environment setting heading.

OBJECTIVES

Anytime a large number of options and variables are in play, the question of optimization arises. Plot betterment, i.e., achieving the best balance between species selection, planting density, and types and

PHOTO 3.2. An agrotechological add-on. In this case, a birdhouse to attract insect-eating birds.

timing of inputs, is a factor in monocropping, more so as the number of variables and directional alternatives expands. Much of agroecology, including temporal and overall farm planning, requires an expanded effort along these lines. This also presents range of economic options and accompanying unresolved issues.

Having a strong sense of direction is essential in agroecology. The leap is not from theory to practice. In agroecology, theory provides abstract reflection that gains strength and direction through implementation concepts, i.e., the agrotechnologies and the design variables. Any economic analysis must further these goals.

As to goals, this is not an easy topic. A short listing of important, but indirect, objectives might include:

- enhancing plot soil fertility;
- efficient input usage (e.g., labor);

- protecting against plant-eating insects;
- keeping weeds at bay;
- moderating extremes in temperate;
- controlling unfavorable water dynamics (too much or too little);
- offering an agreeable micro-climate (e.g., greater humidity, less plant drying);
- avoiding unwanted intrusions or negative impacts on native flora and fauna.

As surrogates, these are strengthened through a more direct rendering. This is:

- increasing plot (land equivalent ratio) LER and/ or lowering costs by operating in accordance with, and not against, natural flora and fauna and nearby natural ecosystems or
- profitability without any unfavorable short- or long-term environmental consequences.

OPTIMIZATION

As shown, profitability is not the end-all. No matter what combination of objectives is sought, achieving these affirms a successful outcome.

Each farmer tries to do the best, given the constraints and objectives. As a section heading, optimization may be too optimistic an expression. Being, in large part, unreachable, the term does capture the direction of the agroecological process.

The problem is that farms and farm plots differ greatly in physical presence and physical characteristics (i.e., soils types, moisture content, topography, etc.). Of more immediate concern, the yields from different crops and crop varieties are at the mercy of the weather and other forces of nature. As constraints, these further complicate the process.

Evolving constraints is one obstacle. The objectives themselves also may be in flux. This can be due to changing internal farm socioeconomics (e.g., a growing family) or the objectives may remain solid, but the markets or other forces that underwrite these may shift.

No matter what happens, at the plot or farm level, optimization is the goal. There is good reason for this, i.e., it is better to fall short of full optimization through an inclusive analysis than to merely drift from point to point without any plan for achieving the best possible design.

ENDNOTES

1. Expressing ecological complexity through a series of underlying mechanisms is presented in Grimm *et al.* (2005).

2. As an aside, vectors do explain broad agricultural policy, that set informally throughout the world. The main thrust, green revolution, is based on the base monoculture, adding ex-farm and genetic vectors. This is not the only policy direction, many more are possible. Examples might include abandoning the pure monoculture for agrobiodiversity and adding a microbial vector or utilizing biodiversity with the rotation and cross-plot vectors. Each combination of vectors could rise to the level of a feasible policy alternative if followed through with a comprehensive program of research.

3. A discussion of domestication, including wild lentils (and the problem of few seeds per stalk and a low germination rate), is found in Weiss *et al.* (2006).

4. The drawback of the green revolution model is shown when droughts intercede. With this two vector approach, there are only two agrobionomic counters: (1) employing a crop genotype that tolerates water shortfalls and (2) irrigation. If neither is feasible, other vectors offer far more drought-countering tools.

5. A listing of 600 wheat varieties can be found in Percival (1922).

6. The varieties of *Brassica oleracea* may be outwardly different but, for other crops, the varietal vector is slightly looser than scientific classification suggests. For rice, most of the varieties are associated with *Oryza sativa*, a few with *O. glaberrima*. In contrast, wheat has numerous species, e.g., *Triticum dicoccum*, *T. orientale*, *T. durum*, etc. Here, as with rice, the varieties also cross species lines.

7. The 40% rice yield improvement figure is from Yoon (2000).

8. For classification purposes, composting would center on the ex-farm (using environmental benign inputs), microbial and the environmental setting vectors. This is in contrast with the green revolution approach which emphasizes the ex-farm and genetic vectors.

9. The manure/ aphid case is from Morales *et al.* (2001), supporting research by Brown and Tworkoski (2004).

10. The notion of path driven optimization may prefigure a distinct branch of agroecological economics. There is no current development along this line.

4 Agrobiodiversity

Nature does just fine allowing multiple species, plants often accompanied by appropriate fauna, to inhabit an area. The same good things are expected with agricultural ecosystems. Multiple plant species; all contributing to a favorable economic outcome, is a form of biodiversity and an exploitable vector.

To achieve success, the agrobiodiversity vector requires different limiting resources, complementarity in plant nutrient profiles, and niches for dissimilar, but densely spaced plants. Success is measured through strong land equivalent ratio (LER) values, lack thereof comes when adjacent plants are overly competitive and the LER falls below one. Keeping the LER high requires resource use efficiency where essential resources, especially those that are limiting, are not left untended or underutilized. Instead, these are exploited to their fullest.[1]

Doing so requires systems so formulated. A number of agrotechnologies have been identified as being resource-use efficient while effectuating multiple outputs. These are the implementation backbone of the agrodiversity vector.

APPLICATIONS

There are many types of omni-productive systems. Of these, seasonal intercropping is most conspicuous. Additionally, there are multi-species plantations, as found in forestry and agroforestry, that offer multiple plant species with multiple outputs. These share many of the same issues and concerns as encountered in intercropping.

With all members of the agroecosystem providing a useful output, there is a need to economically differentiate the outputs. For this, those of the primary species are the most valued.[2] Those offering lesser-valued products are the secondary species. An agroecosystem may contain one or more secondary species, occasionally, two primary species may coexist.

The most prominent, if not the most common use of agrodiversity is purposeful, planned, and managed. The other forms of agrodiversity, expanded, agro-enrichment, casual, and supplementary, adhere to the same principles except that non-planned additions, e.g., useful weeds, are tolerated. Although less noted, these have significance as these dominate on farms in diverse regions of the world. They are as valid as the fully planned, archetype version.

UNDERLYING BIODYNAMICS

This section continues, and provides more details on, the progression found in Chapter 2, i.e., limiting resource, nutrient profile, and exploitable niche differences. This section looks closely at the underlying biodynamics,[3] mostly found with, but not exclusive, in archetype agrobiodiversity.

A number of gains should occur in agrodiverse systems. The ideal mix of productive plants, seldom achieved, is expected to:

- directly increase the plot LER,
- provide protection against plant-consuming insects,
- keep weeds at bay,
- offer a more favorable micro-climate (e.g., greater humidity and less plant drying),
- counter extremes in temperate,
- enhance water inflation and reduce surface runoff,
- halt the spread of plant diseases.

Yield Gains

There are two mechanisms that underlie positive plant-on-plant interactions and produce LER values greater than 1, these are (1) interception and (2) conversion. A number of sub-mechanisms underwrite each. The breakdown is:

Competitive acquisition (interception gains)
- Temporal partitioning
- Exclusive access
- Separate sources

Competitive partitioning (conversion gains)
- Single resource gains
- Multiple resource gains

Resource removal
Absolute gains
Harvest index
Fruit quality

Additional mechanisms may complete the picture. As facilitation and competitive exclusion are generally not a large agrobiodiversity influence, these are presented in the next chapter under facilitation. However, these can and do impact agrobiodiversity decisions and, when full incorporated, these complete the agrobiodynamic picture.

Competitive Acquisition (Interception Gains)

Often two species, planted in close proximity, utilize a single resource more efficiently. These species can share the same limiting resource. Non-competitiveness can be due to improved acquisition (e.g., roots in different soil strata).

Some seasonal species, e.g., carrots, turnips, and beets, have deeper roots whereas crops, such as wheat and rye, are surface rooted. Paired according to their root strata, these can divide water and nutrients in a mutually favorable way. Paulownia (*Paulownia elongata*) is a deep-rooted tree that allows surface-rooted crops to occupy the surface soil with little belowground interference.[4]

Other biodynamic mechanisms come about when competing plants have separate sources of essential resources and, through different resource-use profiles. Often these species do not have the same limiting resources. This can be utilized, through species selection, for competitive gain.

Some plants are more suited to, and seek, vertical (midday) sunlight, others want horizontal (early morning, late afternoon) light. Plants that have vertically tall canopies tend to be seekers of horizontal light, so do plants with vertically positioned leaves. Plants with flat, horizontal canopies look for vertical light as do plants with horizontally inclined leaves. The corollary, one species seeking vertical light paired with a species that looks for vertical light can result in light-use efficiency if they are well positioned.[5]

For light, a visual inspection can tell, to a fair degree, if this resource is utilized efficiently. If direct sunlight strikes the ground, this resource is being wasted. Density and diversity, as well as canopy shape, are often key to efficient collection.[6]

Nutrient profiles are part of separate source biodynamics. Take the case of one species demanding more nitrogen, the other more phosphorus, if the site is well endowed with each resource, they have less of a competitive relationship. This can also happen where two species each want the same resource, but find it through separate sources. Nitrogen-fixing plants find this resource in the air (N_2), a companion species look toward in-ground sources (e.g., NO_2). Barley interplanted with field pea seeks nitrogen in this way.[7]

Part of separate source gains occur when a nutrient is locked away, either physically trapped in rocks or chemically bound with other elements. Gains occur when one of the plant species can extract the trapped mineral resource.

Another biodynamic principle is temporal partitioning, the component species demand different levels of essential resources at differing times. The expectation is for an above-one LER.

A common example, one already cited, lies with the space-constrained backyard gardener. Lettuce in often intercropped with tomatoes where, early in the season, space and essential resources are available to the short-duration lettuce. The lettuce will have been harvested when the tomatoes begin to take the full complement of site resources (Photo 3.1).

Competitive Partitioning (Conversion Gains)

One principle behind conversion gains come from the margin efficiency inherent with a strong essential resource base. This is due to nonlinearity of the resource-yield curves and the marginal gains obtainable, i.e., where, two species, both competing for a single resource, can result in a favorable LER.[8]

Figure 4.1 illustrates two hypothetical plant species, growing together, competing for the same resources. The division of this limiting resources, 50% going to each species, results in a LER that exceeds one. Figure 4.1 shows an LER of 1.5 (0.8+0.7).

A second conversion-favorable principle happens when a resource is removed. Seemingly counterproductive, not all mechanisms rely upon gains through added essential resources. A reduction, caused by one plant in the amount of one or more resource going to a second species, can increase the overall productivity. This underlying mechanisms is often expressed through the harvest index or in an increased quality of the output.

Absolute gains occur through overabundance. Too much light can cause excessive drying and moisture stress. Water-soaked soils can be similarly anti-productive to many crops.

In both these cases, a second species, one that appropriates or soaks up the overabundance, can be a well-planned addition.

Harvest index gains also occur through removal of a resource. Certain resources may upgrade one part of a plant, e.g., light can promote foliage over fruits. The harvest index, as the ratio of biomass to output, is a measure of this. This has been shown to be true where a

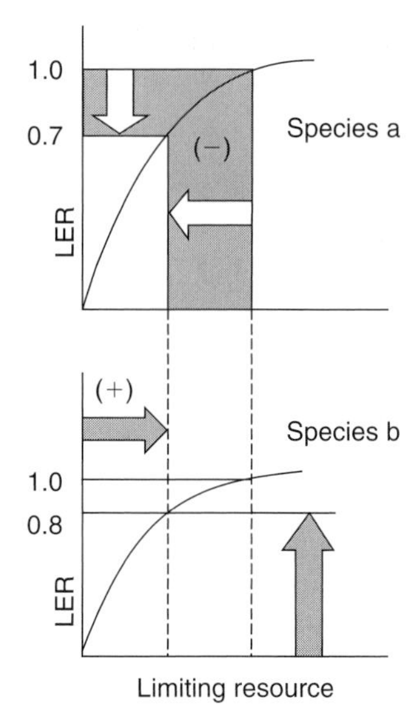

FIGURE 4.1. Marginal gains where one half of the limiting resource is taken from species a and allocated to species b. Rather than half the potential productivity from each species, marginal gains allows for a LER of 1.5 (0.8 + 0.7).

millet–groundnut intercrop had a LER of 1.2 for biomass alone, an LER of 1.33 for marketable outputs.[9]

Gains in fruit quality are well documented. This can be a case where yields are smaller, but quality of the output, and may be the selling price, improved. It has been shown that the protein content of grains are higher, the taste of coffee better, and, for trees, the amount of the more decorative and decay resistant heartwood is greater if subject to interplant light competition.[10]

Cost Gains

Some clear cost savings occur when two species share the same plot of land. One cost gain, the weed control possibilities for intercrops, is well established. The first mechanism is in the number of soil niches occupied. Weeds seek space and essential resources and, if these are engaged, set groups of weeds fail to thrive.

A second mechanism lies in choosing a plant species that suppress weeds. Some are particularly good at this. Large-leafed, ground-hugging plants seem to excel, examples being plants of the squash family.

It may be the biomass, particularly the contained chemicals, that acts as a weed suppressant. This can be on-site or the weed-inhibiting biomass can be cut and carried to the location and spread around the base of the crop plants. The cut stems of barley, oats, and wheat do suppress weeds.[11]

Risk Reductions

Clearly, the failure of one crop in an intercrop is a loss. Still, if others in the mix yield, the system is not an entire failure. There are a number of ways this is done in an agrodiversity context.

Climate

Intercrops counter climate through standard means. A dense mass of vegetation does produce a localized micro-climate, i.e., daily temperate extremes are moderated, the internal humidity can be higher and per plant transpiration lower within a crop plot. The result, less overall water use.

There are other anti-risk mechanisms, some more direct. A wheat field in a desert setting can provide grain in good years, become a pasture when rains are less than sufficient for a grain harvest. Another rainfall-related option co-plants a drought-resistant and a water-demanding species. In high rainfall years, both will prosper and yield. In low rainfall years, the drought-resistant species will yield, the second offers little except for facilitative services.

Insect and Plant Diseases

It is well documented that plants diseases and some herbivore insects prefer an assured micro-climate. By providing something else, diseases will not inflect, certain insects populations will not thrive. This applied to tillage and types of fertilizers used, some aid, some dissuade in-soil or on-soil micro flora and fauna.

It is well known that flourishing plant populations, those with plentiful essential resources, better fight attacking organisms. It has been demonstrated that fewer pines per area, more per plant essential resources, and healthier trees confers resistance to pine beetles.[12]

Rather than rely on the community and management inputs alone, plants can take more of a facilitative role. Many plants repulse insects, some are better at thwarting certain species than others. In a partial listing below, given are a few plants particularly good at opposing one or more specific insect pests. These can be a useful intercropping addition[13]:

- Chives (repels aphids and mites)
- Geranium (leafhoppers)
- Horseradish (Colorado potato beetles)

- Rue (Japanese beetles)
- Sage (cabbage maggots and cabbage moths)
- Spearmint (ants and aphids)
- Thyme (cabbage moths)

Relevant Guidelines

For the above, where soils are relatively well endowed with water and a range of nutrients, two well-chosen yielding species (as through agrobiodiversity) can exploit the essential plant resource base better than a single species. When soils are less well endowed, a facilitative plant-on-plant relationships (as through directed biodiversity) may yield better. This is a key guideline in the selection of companion species.

When a site is fertilized and irrigated, agrobiodiversity with resource compatible species many be the first option considered as this will almost always result in a high LER. Costs, not revenue, may override the LER gains, i.e., where the cost of hand harvesting the different outputs may far exceed the cost of machine harvesting a monoculture.

Beyond this, other rules take hold. For either agrobiodiversity or biodiversity, the companion species, planted in close proximity, is nearly always the best spatial option provided these coexist well. These are less density planted when the companion species are slightly competitive for one or more essential resources.

It is not uncommon to find a mix of competitive partitioning and facilitation where some essential resources are divided to advantage, others enhanced through a facilitative relationship.

The matching of plants, through their inherent characteristics, is one means to achieve productively favorable pairings. Amplifying this theme, some general guidelines can be proposed for the close pairing of a slow with a fast growing species.[14] These are that:

1. A light-demanding species is interplanted with one that is shade tolerant if the light demander grows faster.
2. A slow growing light demander is only intermixed with a faster shade tolerant species if
   (a) outside help is offered, e.g., pruning, thinning, or a less dense planting;[15]
   (b) clumps or strips (where like plants are in groups) are used instead of fine patterns (see Figure 4.2); or
   (c) planting is temporally staggered which allows the slower species a competitive footing.
3. When combining two or more shade-resistant species, the growth rates should equate or the slower one is protected, through pruning, thinning, or spacing, from being overwhelmed.

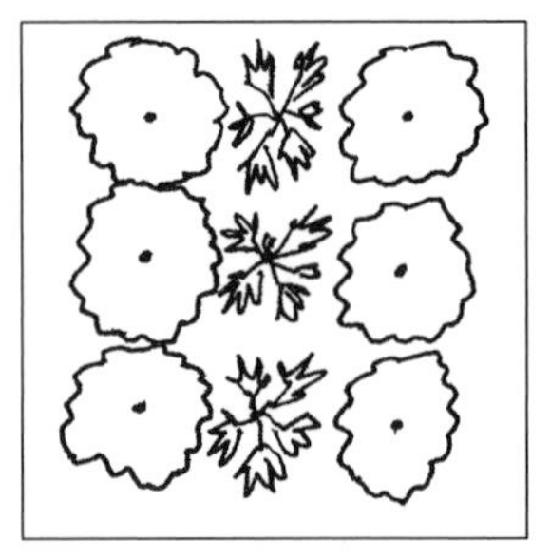

FIGURE 4.2. An overview of two fine patterns where each plant is in direct contact with an unlike species. With favorable plant–plant complementarity, fine patterns have the potential to produce the highest LERs. On the left is a checkerboard pattern, the right is a row pattern.

4. Two or more light-demanding species should not be admixed, except
   (a) on very favorable, i.e., fertile, well-watered sites;
   (b) where one or more of the light demanders are harvested (removed) early in the growing season.

These guidelines apply, with caveat and exceptions, to relative height growth. In addition to species planted at the same time, the first rule can refer when a tall perennial is paired with a short annual. The operative exception is when a taller plant with an open, light admitting canopy is placed over a shorter, shade-tolerant species.

Another set of guidelines exists for vertical and horizontal light gathering. Addressed through spatial patterns, those that seek horizontal light are placed around the perimeter of those that want vertical light. Another option refers to row orientation. Shorter plants that desire vertical light are placed in north–south strips between single or double rows horizontal light seeking plants. The idea being that, at noon, direct sunlight illuminates and encourages the shorter plants.[16]

Less can be said for belowground interactions. Much remains toward formulating a concise, coherent set of applicable ground (i.e., belowground) rules.

Although subjugated by numerous caveats and exceptions, incomplete guidelines still serve a practical purpose, they focus thought by trimming, to manageable proportions, an extremely long list of possible species pairings and variables (such as planting densities). In predicting when and where plant pairings can best succeed, future guidelines, expanded to encompassing design variables, could put agrobiodiversity on a less subjective foundation.

ECONOMIC MEASURES

The LER is a measure of site-based essential resource efficiency well suited to agrobiodiversity.[17] This is both intuitive and intimate with in-field happenings. The lack of monetary units in the calculation keep the focus on productivity, not on the selling price. The immediate questions are on planting density, specifically the balance needed to attain the highest LER. The analytical methodology leads in the direction of the production possibilities curve (PPC).

The Production Possibilities Curve

The PPC is a well established means to evaluate intercropping. Although concise in conceptional expression, this seldom finds a practical voice. Still, the technique has merit and is integral toward understanding the economic eventualities.[18]

When used to determine planting density, the problem is that there are a wide array of spatial patterns and arrangements. Finding which is best and at which densities is one use of the PPC. The archetypal PPC is based on raw production data, that presented here utilizes LER values.

Ratio Lines

A simplifying factor in deriving a PPC are ratio lines. For every spatial patterns, there are planting density considerations. These can be expressed as planting ratios. For the monoculture, there can be a single-recognized planting density that is considered optimal.[19] This is a bit more complex when more than one plant species contributes to the economic outcome.

The normal, optimal, monocultural planting densities underlie the end point of the central or midpoint line. This ratio is 50–50%.

Take the case where the optimal planting density for a monoculture of species a is 750 plants per area. Similarly, the optimal monocultural density for species b is 3000 plants per area. Taken together, this is the outer most point along the 50–50%, so marked as the upper right point in Figure 4.3.

Outward along this ratio line lies a series of joint planting densities. In this case, some points along this line are 500–2000, 750–3000, or 1000–4000 plants per area. To make a PPC, more ratio lines are needed. Figure 4.3 also contains lines for the ratios 75–25% and 25–75%. Others are possible.

Curve Derivations

In Figure 4.4, a third dimension has been added to Figure 4.3, that of the LER surface. This overlays various planting ratios and densities.

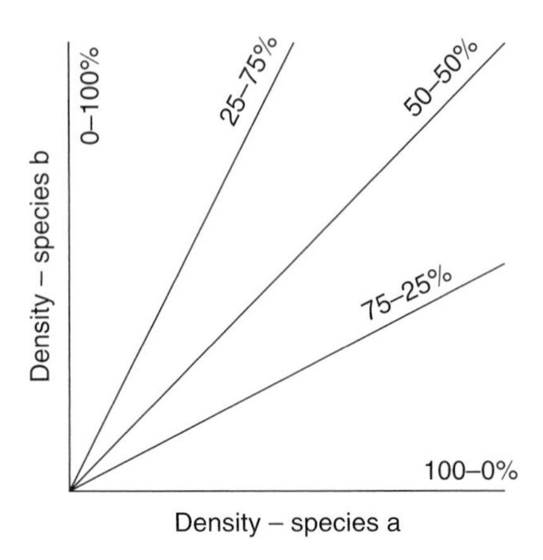

FIGURE 4.3. Individual ratio lines, each of which has the per plot populations of two interplanted species in constant ratio. These underwrite and simplify deriving the PPC.

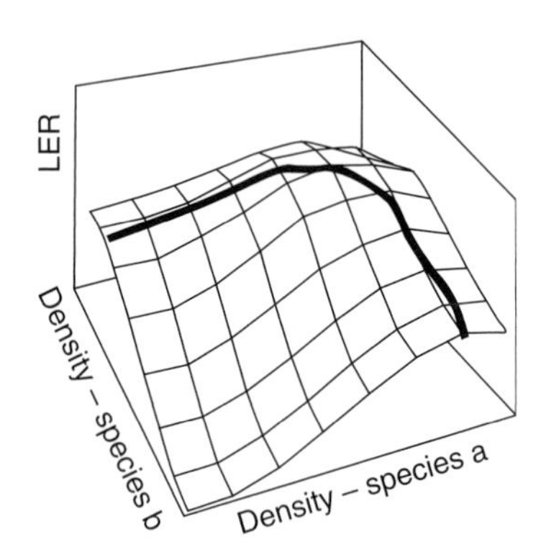

FIGURE 4.4. Derivation of the PPC. The vertical dimension, that expressing the LER at each density combination, overlays Figure 4.3. A line drawn along the ridge of the LER values demarcates the PPC.

A line drawn along the ridge of this LER surface constitutes the PPC. In Figure 4.4, the uppermost point optimizes a spatial pattern as to the interspecies planting ratio and the planting densities.[20]

The normal expression of the PPC is two dimensional (as in Figure 4.5, left).[21] Derived in a slightly different manner, the point along the Figure 4.5 curve, that furthest from the origin, is the highest LER.

In determining the ratio lines and PPCs, many practicalities intercede. For example, when planting and/ or harvests require certain spacial arrangements, not all row patterns and not all possible densities would be included. This is the case when farm machinery forces certain spatial patterns and results in a comparatively narrow range

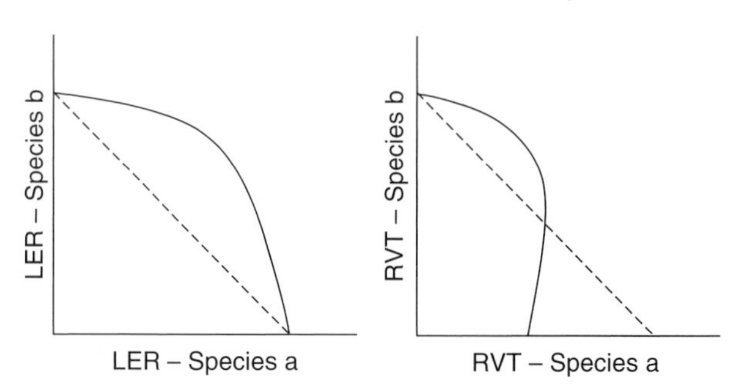

FIGURE 4.5. Two PPCs, one employing LERs, the second uses LERs with market or selling prices included.

of densities. Nonetheless, a well-established curve, with an optimal point, is a good starting point for any pattern or any so derived agrotechnologies.

Price-Adjusted PPCs

There is another complication, the PPC, as stated, assumes that the user wants the best LER irrespective of the market value for the individual crops. Take the maize–bean intercrop, if the selling price of the beans are substantially higher than that of the maize, the farmer may want more beans, but may still desire a functioning mixed species ecosystem with the inherent ecological advantages.

There is a simple solution, the relative value total (RVT) takes the place of the LER. The result is a revenue-adjusted PPC with a different optimal point than with the unadjusted version. Figure 4.5 shows two PPCs, one is unadjusted (Figure 4.5, left), for the other prices have been added (Figure 4.5, right). With the price reformulation, the point on the curve that is further from the origin is the best with regard to joint productivity as tempered with market or selling prices.[22]

Ratios excel at comparison, a ratio-derived PPC being no exception. There is one strong proviso, one must remember that PPCs are pattern dependent. The data for one spatial pattern, say a row layout, cannot be merged with that of another, say an individual design, i.e., one cannot mix the patterns in Figure 4.2 into one PPC. These must be presented as discrete curves.

DESIGN VARIABLES

If no simplifying measures are taken, it is easy to become awash in variables. For straight agrobiodiversity encapsulation, the four alternatives are (1) species, (2) temporal adjustments (3) spatial pattern, and/ or (4) planting density. The 'and/ or,' or cross-effects of these variables makes this a difficult assignment, especially when optimization is expected.

Species

It goes without saying that two distinct species, mutually exhibiting some degree of complementarity, are always the best starting point. Some are known, others can be surmised from the application of the biodynamic principles.

What is asked is the magnitude of interspecies complementarity. High complementarity (e.g., greater than 1.80), on the site or sites in question, makes usage much easier. Complementarity is measured through the unadulterated LER.

Planting Density

The starting point, given known interspecies complementarity, is always the optimal planting density. This applies to each co-planted species, e.g., if a monoculture of species a is planted at 10,000 plants per hectare, species b having, as a monocrop, an optimal density of 15,000 plants per hectare, the intercrop would be composed of 10,000 and 15,000 plants per hectare, respectively.

Maximum density is not the only option, different dimensions can be altered for gain. In one case, a change in interrow distance for maize was recommended as a way to increase the yields of understory soybean. This allowed more light into the understory, increasing the soybean yield and the overall LER.[23]

If more is known on how species a and b will behave together, then appropriate adjustments in their respective densities can be the starting point. The end goal, that which optimized this variable, are actual dimensions, e.g., precise inter and intrarow spacings.

Spatial Patterns

Spatial patterns, in league with planting density, are often the route to site resources use efficiency *via-a-vis* the plant species chosen. The highest LERs are found when the co-residing species are complementary. Favorable interspecies dynamics are often brought to the fore when the paired species are in close contact, as through fine patterns (Figure 4.2). When paired species have less complementarity, other patterns, e.g., blocks, or strips, may best capture the existent interspecies dynamics.

PHOTO 4.1. A backyard garden exhibiting, in miniature, vegetables in a block pattern. At the smallest end of the scale, larger versions some spanning many hectares, e.g., as in Photo 13.2, are possible and common.

FIGURE 4.6. In row cross-section, two agrodiversity formulations. On the left is treerow alley cropping, on the right is a light shade system.

What further complicates the situation is that, within each spatial pattern, the actual inter and intraspecies distances must be set. This can result in numerous pattern variations.

Without going into the details, the simplifying factor is that a basic spatial pattern is a component in the design of an agrotechnology. For example, the row cross-sections in Figure 4.6 are inherent with hedge or treerow systems. There still remains a need to optimize inter and intraspecies planting densities, the outcome of which can significantly influence the LER and resulting the economics.[24]

Temporal Adjustments

Pairing a fast growing species with one that progresses slower can be a recipe for disaster. To extract the full measure of complementarity, minor temporal adjustments can overcome this problem, e.g., delaying the planting of a second species by a few days or weeks or coating the seeds of the second to hold off germination.

In an already cited example, beets with wheat, excessive and negative belowground competition occurs if planted at the same time. The preferred alternative is to first plant the root crop, allow the roots to reach into the lower strata, then plant the surface-rooted wheat. This entails a delay of about 2 weeks.[25]

Beets are notably shade resistant, so over topping by the wheat is not an issue. Also possible are onions or cabbage with wheat. Cabbage can shove aside the wheat stalks in the quest for light, onions, with a growth jump, can compete favorably for same horizontal light.

As for the wheat, this is not considered a marketable crop, it will be trampled when the primary species is harvested. As a facilitative species, the gains come in suppressing weeds, blanketing the ground during the off-season, for in-place forage, and as a decoy crop for grain-loving birds.

In finding the optimal bicultural delay, simple LER analysis works well. Additionally, there is the option of employing a modified PPC. The theory being that an optimal delay puts both species on an even footing, a lesser delay favors faster growing plants, a longer than optimal delay benefits the slower growing species.

Density, Diversity, Disarray, and Duration

For the ecological dynamics of complex agroecosystems, natural or human influenced, to thrive, certain parameters must be in place. These are density, diversity, disarray, and duration (also phrased as biodensity, biodiversity, biodisarray, and bioduration).

Density insures that each component species is in close enough contact so that multitudinous plant–plant interactions occur. The generally rule holds that the closer the spacing the better the result. This comes with caveats. The climate and essential resource must support a large number of plants in a small area. The high density rule holds, and is not over stretched, knowing that a thriving ecosystem can overcome considerable climatic adversity, even situations that would doom the individual plants.[26] There is practical concern, room, often as paths, is needed for harvests.

The gains from biodiversity are well stated. The complex system, through a more-the-merrier approach, takes this to the extreme. Observation and formal study puts the diversity crossover line at about eight species, above which an expanded array of natural dynamics take hold.

Disarray insures that the component species are spatially arranged such that plant-on-plant interactions are approaching some maximization point. The best scenario may be when one species is in contact with as many other species as possible, promoting as many types of interactions as possible.

A random placement can work well, but a pattern may be better. Often disarray comes with an underlying pattern where like species are seldom clumped, but spaced (at varying distances). To promote the best possible biodynamics, users help by placing individual species where they think they will do best.

Duration assumes that an agroecosystem exists long enough so that there is seasonal or yearly carryover of living organisms, i.e., the large diversity and large populations if micro and macro flora and fauna do not have to begin again anew following each planting. This does not mean that ecosystem must be perennial, only that there is enough of an ecological continuation from one season to ecologically jump start the next.

Complex ecosystems without density, diversity, disarray and duration, are difficult to manage through plant–plant governance. The better course is to allow nature a free hand, through ecosystem governance. This shortcuts the need for spacing, temporal information, and a planning regime (with set row and interplant distances).

The resulting ecosystems most often have very good environmental properties and recognized high overall productivity. These are effective when the user accepts many, but not much of any one output. Under this scenario, economic performance is a tangible, but mostly nonquantifiable.[27]

The Non-harvest Option

There are times when one or more of the outputs from a multi-output system are not harvested or only a portion is taken. This is the non-harvest option. This may be the strategy when a selling price plunges making one of the harvests financially unattractive.

There is more to this. A producing plant can be an intercrop component more for it's facilitative benefits than as a revenue source. When this happens, the non-harvest may be the underlying strategy except where a market or use appears, making the output more of interest. The addition of squash or pumpkin to a maize–bean intercrop can be a non-harvest situation. Given the expected low yields of the squash and the cost reduction gains in the form of weed control, the high per unit cost of a harvest may mean abandoning, i.e., not harvesting, this component.

The non-harvest option can be utilized in risk containment. A primary species can be climate sensitive, i.e., when rainfall is below the norm, yields fall dramatically. When this occurs, the secondary species, one that is more resistant to water shortfalls, is planted. When rainfalls are adequate, the accompanying secondary species would be harvested. If rainfalls are low and the primary species does not yield well, this output is not harvested. The secondary species fills the void as a natural form of crop failure insurance.

ASSOCIATED AGROTECHNOLOGIES

The agrobiodiversity technologies have in common that all or most of the purposely placed species offer an economically interesting output. Inclusive in this are versions of the monoculture. This may not preclude a few non-productive, facilitative species in the more species diverse versions.

Listed, the agrotechnologies groups under an agrobiodiversity vector are:

- Monocultural
  - Pure
  - Varietal/ genus
- Productive intercropping
  - Simple mixes
  - Strip cropping (seasonal)
  - Barrier or boundary
  - Complex agroecosystems (without trees)
- Productive agroforestry
  - Isolated tree
  - Alley cropping (treerow)
  - Strip cropping (mixed tree)
  - Agroforestry intercropping
  - Shade systems (light)
  - Agroforests

The above list is, in some ways, incomplete. The focus is on agrobiodiversity and those systems which contain, by intent, all productive species. All, except for monocultures, can be reformulated with one or more facilitative species. The same holds true with biodiversity technologies (next chapter) where a facilitative species can be replaced with one that is both facilitative and productive.

The majority of these agrotechnologies, in seeking a higher LER at increased costs, are revenue oriented. As noted, there are exceptions.

Cost and revenue orientation are part of the initial weeding out of those systems deemed unsuitable for a specific application. Most often only one or two agrotechnologies will be best suited for the socioeconomics of a farm or farm plot. The parameters of use that accompany each agrotechnological description (as provided in this and subsequent chapters) can be the sole basis for initial implementation decisions.

Monocultural Variations

Although seemingly one system, the monoculture does offer variation, enough so this establishes a category of, not a single agrotechnology. The root form is the pure, or genetically near-pure, monoculture. Other types introduce more genetic variety.

Pure Monocultures

As the base of vector notation, pure monocultures are widespread. This are based on a genetically improved species, armored to resist threats, augmented by ex-farm inputs. The simplicity is enticing and the economic possibilities well established, however, these come with a well-deserved reputation for being ecologically out of sync in all regards.

Varietal/ Genus Monocultures

Multiple varieties of the same species, when intermixed, can provide some of the benefits of intercropping while still qualifying as a monocropping. Some go a bit further, rather than relying upon varieties of the same species, mixing plants within the same genus.[28] The principle advantage of a varietal or genus monocrop is in the protection offered against harmful insects and plant diseases. Documented examples include rice in China, wheat in Pakistan and Syria, and maize in Mexico.[29]

For undifferentiated products, e.g., animal feeds and sugar or cellulose for ethanol,[30] a varietal or genus based monoculture may be the best course. For most applications, the key economic element, and obstacle, lies in being able to mix and co-market that produced, e.g., grains, each with a slightly different look, feel, and/ or taste. For vegetables, separating the outputs may be the least of the concerns, finding markets for different varieties of the same species, given consumer uncertainly, can make these intraspecies or intragenus mixes problematic (Photo 4.2).

Productive Intercropping (Non-woody)

Productive mixes are often though of as seasonal and, in part, this is true. There are intercrops that do not have to be replanted; these spring

PHOTO 4.2. A monoculture of intermixed lettuce varieties.

anew from root stock or naturally re-seed, most are non-wood perennials but some shrub-type woody perennials also qualify. The agrotechnologies presented in this section are for non-woody plants.

Simple Mixes

When plants exhibit cross-species complementarity with another cropping species, a high LER is possible. Being the main economic goal for simple mixes, complementarity is best accomplished when two or more species are densely interplanted. The maize/ bean intercrop is the classic case where high densities are the rule. However, there may be cost considerations, a maize/ bean/ squash intercrop includes the squash, not altogether for the output, this is secondary, but, as previously stated, for increased weed control.[31]

This category does not lack representation. Vandermeer (1989) starts by listing 45 different seasonal bicultures. On this list are peas with oats, soybean with rice, and maize with sweet potato. Considering that this is only a small window onto a large and mostly unstudied topic, this tally of simple pairings would be low. However, the large number found does provide indirect proof into some favorable economics.

Although these present harvest difficulties, simple mixes, because of the high degree of land-use efficiency, are appealing, especially where landusers are boundary constrained and have a large labor-to-land ratio. As a result, these are most often high input, high output, found in or near population centers and/ or where overpopulation keeps farms small, the food and revenue needs high.[32]

Strip Cropping (Seasonal)

Taking advantage of terrain, strips of productive plants, often contouring hillsides, provide erosion control and increased capture of water runoff. An individual strip can be monocultural or contain some intercrop. Because of the large amount of inter-strip interface, it is better if non-competitive species are in adjacent strips. This is less a requirement if the strips are broad (as in mechanized commercial farms), more so if the strips are only a few rows wide (as found with intense subsistence farms).

Rather than worry about the competitive tendencies of adjacent crops, there is the option of planting an between-strip buffer species.[33] The intent is to reduce inter-strip competition by stopping the crop roots from growing under adjoining strips. Usually a perennial plant with deep, vertical roots, buffer species allow for greater cropping flexibility in that the crops in neighboring strips do not have to be paired for growth complementarity.

With or without a buffer species, the key is to have plenty of cropping diversity across the system. The ecological spillover from the component species can halt the spread of harmful insects and diseases again while aiding with erosion control.

The environmental advantages and long list of ecological gains, plus the ability to engage farm machinery, makes this an attractive, broad application agrotechnology. Issues do remain, but the economics should generally be pleasing. As this is visually different from the spacious monocultures of large commercial farms, it is the intangibles, including a reluctance to take the initiative, that may prove the main obstacle to widespread adoption.

Barrier or Boundary

It is not unusual for landusers to position a barrier or ring plots with some plant species. Productive plants can serve well. One reason may

PHOTO 4.3. An exposed garden plot located at the corner of a large pasture. This is protected at the periphery by rows of Jerusalem artichoke.

be micro-climatic, a tall boundary will block drying winds. Examples of yielding, and wind blocking, boundary species are maize, sunflower, sugarcane, and Jerusalem artichoke (the latter is shown in Photo 4.3).

Other functional mechanisms include acting as a decoy crop, directing birds from the primary crop or helping to curb the spread of unwanted insects. In one of many documented cases, buckwheat around cucumber protects from the cucumber beetle.[34] Barrier plants differ from strip buffers in that the interactions are mostly aboveground.

The question is which species to use and if the economic gains outweight the increased management involved. Given the range of gains and the number regional applications, most of these questions have been satisfactorily answered, as least for small farms. For larger commercial applications, barrier species tend to be non-productive (as described in the next chapter).

Complex Agroecosystems (Without Trees)

Mixing a large number of season crops does insure at least some of the ecological gains from a dense, diverse, and disarrayed plot design. For seasonal species, the gains from duration may be less obvious, however, positive effects do linger in the soil and can carry across seasonally as long as an positive environmental setting is continually maintained (e.g., no bare land phase, only unabated cropping). The result, these are more of a humid tropical agrotechnology.

Like simple mixes, these may remain a landscape feature where the land area is constrained, markets for diverse produce are good, and labor is relatively abundant. Outside these conditions, the economics may be far less favorable. Because it is operationally cumbersome to intersperse so much in a large-scale enterprise, these remain more of a household or a backyard undertaking.

Productive Agroforestry

Many think of agriculture as more seasonal crops. The concepts presented do apply to orchards and tree plantations, trees in monoculture or as more elaborate mixes. Some utilize a few perennial tree species with season cropping, other systems are entirely tree based. This category is very diversity in purpose and economic outlook.

Isolated Tree

The isolated tree system is oftentimes a large fruit tree standing alone in a spacious plot. A large tree, fruiting or not, may provide a habitat or a resting place for birds that feed on crop-eating insects or crop-damaging rodents. There might also be some gains in wind protection.

As these are seldom pruned, isolated trees, if not plot dominating, are not an economic issue of any note. In a large plot, the crop losses from one or two trees in a large area are trivial, the ecological gains are also correspondingly small. The use decision is mostly based around the intangibles; a shady place for workers to rest, some fruit outputs, esthetics, etc. There is no major limits on isolated trees systems except where highly mechanized operations find the stems inconvenient.

For grazing systems, the economic importance may be greater. Shade and shelter can help cattle or other domestic livestock gain weight and increase the survival of the young. Such trees, with or without fruit, can be a substantial economic addition.[35]

Alley Cropping (Treerow)

It is quite possible to plant crops between rows of trees. For this, crops strip are between single rows of trees. For this to succeed, the canopy of the tree must not expand laterally to any degree, e.g., with dwarf palms and dwarf fruit trees. The other option is to prune the trees so that the crops get sufficient light. The defining characteristic is the treerows and the open space above strips of light-seeking crops. The trees only touch within each treerow (as in Figure 5.4, right).

As the tree fruits and other products provide the greater share of the income, this will be the primary crop. To avoid trampling the lower crop, the understory must be harvested well before the treecrop. This requires an early maturing seasonal understory. Root compatibility is less a factor as preseason plowing will prune any surface roots that spread into the crop strip.

Treerow alley systems seem to be most efficient when the trees gather horizontal (early morning and late afternoon) light and the crop is allocated vertical (noon) light. These require a north–south row orientation. This may be ecologically inconvenient as wind movement and hillside contouring are interceding variables. Because of the row orientation issue, the few treerow alley systems encountered are mostly on sheltered or wind protected flatlands.

As above, it is the exacting requirements (i.e., a suitable tree and crop species, the harvest timing, as well as a level location) of these revenue-oriented systems that make these less than common. If the requirements are met, these are a sure economic addition.

Strip Cropping (Mixed Tree)

Rather than single rows of trees alternating with crop strips, there is the option of trees planted in strips. The dynamics are much the same as single row systems except that a wider area is taken by the trees. The treecrop from the perennial tree strip can be of fruits, nuts, or some tree product. This allows two options on how the treecrop is harvested.

For the first, the tree strip is internally gathered. This provides a bit more harvest flexibility than their single row counterparts, i.e., pickers do not have to tread on the crop strip/ alley to collect the treecrop and allows for annual or perennial crops between the trees.

The second option utilizes the crop strip both for a yielding crop and to reduce the cost of a high-volume treecrop harvest. This is done by allowing trucks and other machinery onto the crop strip. The main proviso has the crop being harvested before the fruit.

The drawbacks are the same as with single-tree systems, the crop alley must be kept free of overhanging trees and the need for light efficiency and north–south row orientation often requires a sheltered, flatland location. If shade tolerant crops, either annual or perennial, are utilized, light efficiency through row orientation is less a factor and the strips can contour hillsides. An example would have shade-resistant coffee alternating with strips of fruit trees. The lack of experience with these designs limits the application possibilities.

Instead of fruiting trees, trees can be planted that are harvested when young, e.g., Christmas trees, cinnamon (the bark being destructively harvested), small palms felled for palm hearts, and pole crops. The smaller trees and the early cutting means that light and row orientation are less a factor. For this, seasonal crops have an slight advantage as processing the trees when the crops are not present allocates plenty of work room and can occupy off-season labor. Strip plowing insures that the tree roots will not seek nutrients at the expense of the crop.

Crops strips with tall trees may prove more of interest for their ability to counter frosts rather than for their revenue-orientation possibilities.[36] Systems with frequently harvested trees are constrained simply because so few raise trees for a speedy harvest. Operational issues aside, there is also the issue of having prime land tied up in hard to remove trees. This may limit use to large-scale commercial farms.

Agroforestry Intercropping

Rather than having one or two isolated trees, a user may opt for a fairly substantial fruit or other treecrop presence. There are two versions of agroforestry intercropping, the first might be described as mixed species orchard or treecrop plantation, the second as widely spaced orchard or plantation with a crop understory.

For the first of these, a mix of fruit, nut, or other bearing trees are interplanted in an ordered system. Traditional German streuobst is of this type, i.e., with mixed varieties of apple, pear, cherry, and other fruit trees. In this version, the tree canopies in the mature stage will lightly touch. Pigs and other animals feast on the non-harvested fruit.[37]

The economic gains of mixed treecrops come through risk reductions and in being able to keep harvesting as the different fruits mature. Instead of the single-variety harvest lasting 2 or 3 weeks, with multiple varieties, e.g., apples, this can extent 2 or 3 months. This favors those that market small quantities daily. This also benefits the animals that depend on this fruit.

In contrast, a greater emphasis can be placed on ground-level outputs. What differentiates this second version are the wide tree spacing (as in Figure 5.4, left drawing). The purpose is to allow the crops beneath to receive light without much overstory interception. The spacing and root-trimming from plowing limit cross-species root interference. One example has cucumbers grown underneath widely spaced oranges.[38]

As a revenue-oriented system, a high LER is expected. The other economic concern is with harvest, the understory must be gone before the tree produce is picked. These are not as economically confining as treerow alley systems. e.g., crop need not seek primarily vertical light. Still, to be economically viable, these must operate with the narrow limit set by the crop-first harvest requirement. In their favor, this can be a machine or a hand picked crop.

Shade Systems (Light)

Light shade systems utilize productive trees as an overstory. The trees are usually tall and spaced such that the canopies touch. The idea being that ample light penetrates the canopy, enough to support active and economically rewarding crops. This require trees with sparse, open canopies (Figure 4.6).

Palms have open, non-spreading canopies and are commonly found in these systems. The coconut palm seems the universal favorite. Also encountered as overstory species are fig, avocado, breadfruit, or jackfruit. These must be widely spaced so that ample light reaches ground level.

The understory tends to be perennial species. Among the wide ranging choices are black pepper, vanilla, pineapples, tea, cocoa, or coffee. Grasses and grazing is another preferred ground-level crop.[39]

These are LER, hence, revenue-oriented systems. As the fruits of the overstory must be harvested, space must be provided for access. The common pattern allows one strip for the crop, the next is kept crop free to harvest the treecrop. This reduces the area for the understory. With these systems, it can be hard to determine the primary species, the result, the overstory and understory can be co-primary species.

As light shade systems are less encountered, the economics seem more on the margin. One often overlooked gain lies in weed control, there being fewer unwanted plants under two canopies (the over and under stories both intercept light).

Not to be forgotten are the intangibles. These provide a cool place to work in the hot afternoon and they do have esthetic purpose. As a diverse ecosystem, birds find these a haven, especially when nearby natural forests are lacking. However, with multiple outputs and the presence of fruit-eating bats and birds, wildlife may be less of an ally. On the positive side, these systems filter and hold rainfall, if the area enveloped is large enough, this will mitigate floods and droughts.

Despite the intangibles, it is the management, costs, and the inconvenience that may tilt the decision process against large-scale commercial use. There are exceptions for specific crops, but generally light shade systems are found more on smaller farms with the labor to adopt revenue-oriented systems.

Agroforests

Agroforests are a large grouping alternatively referred to as tropical homegardens, agricultural forests, or forest gardens. These terms underlie some categorical divisions, some produce foodstuffs for households (homegardens), the products for others are market based (forest

FIGURE 4.7. A comparison of two complex agroecosystems, each formulated around biodensity, biodiversity, and biodisarray. On the left is a single story complex agroecosystem without trees. A multi-story agroforest is on the right.

gardens), or offer a mix of agricultural and forestry outputs (agricultural forests). Avoiding a protracted discussion, the variants of the agroforest are presented in Chapter 14.

These systems are based around density, diversity, disarray, and duration and, because so few outside inputs are required, these are ecosystem governed. Visually, these are often mistaken for natural forests. For this good reason, the same parameters, density, diversity, and disarray, exist with the dynamic of natural forests. The difference is that agroforests are productive units containing predominately productive species (Figure 4.7).

Agroforests offer outputs in large number, 10 is at the low end of the range, the top can offer well over 100 species. The number and volume of output from each is less an economic problem, more one of management. Since agroforests are, in large part, a internally dynamic, self- or nature-governed, stand-alone units without need for research on the best species mix, the economic question is more yes-or-no; to utilize or not.

These systems are exceptional in other aspects. Despite the cornucopia of outputs, agroforests are mostly cost oriented.

Any yes decision will be reenforced by a slew of non-quantifiable intangibles. The contribution of active agroforests to local diets can be difficult to ascertain, especially since much is gathered for immediate use and does not go through local markets. Other incalculables have agroforests as storehouses of biodiversity. Over 325 plant species have been found in a single agroforest. Also in abundance are native birds and other wildlife that, without the agroforest, would not survive.[40] Additionally, there are the intuitive and incalculable benefits of water conservation and climatic modification when agroforests are pervasive across rural landscapes.[41]

ENDNOTES

1. A discussion of agrodiversity, as found in Brookfield (2002), diverges slightly from this.

2. Referencing the RVT equation (2.3), the denominator of the RVT is always the primary species.

3. The words biodynamics or agrobiodynamics are synonymous with agrobionomics (Wojtkowski, 2002).

4. Root strata research for common crops is not that common, one study is by Weaver (1919). Yin and He (1997) and Wang and Shogren (1992) discuss paulownia root architecture.

5. More on light gathering is found in Beets (1982, p. 54) and Donald (1963).

6. Beer's law is the basis for light interception in monocrops. This has been expanded to include intercrops (Keating and Carberry, 1993).

7. Barley with field peas is from Poggio (2005).

8. The conversion or marginal gains are, for well-watered, nutrient-rich sites, may be the primary mechanism of LER advantage. This is unproven, but a logical offshoot. The resulting axiom; when planning for rich sites, intercropping should be the first alternative considered.

Going further, all the other mechanisms of conversion gain may be expressible as an outgrowth of margin gains. Although work along these lines is lacking, this has implication when mechanistically evaluating why productive (LER) gains occur in multiple output situations.

9. The improved harvest index from biodiversity comes from Ong *et al.* (1996). The example, millet/ groundnut, is from Ong (1991).

10. Quality improvement in grains is from Nandal and Bisla (1995) and, for coffee, Muschler (2001).

11. Examples of weed-inhibiting mulch are found in Kamara *et al.* (2000), Chou (1992), and Jobidon (1992).

12. Healthy pines providing resistance against beetles is from Waring and Pitman (1985).

13. Lists of repellant plants are common in organic gardening books. A more detail listing, including the intercrops, the insect species (one or more) controlled, and the stated mechanism, is found in Altieri and Nicholls (2004, p. 80).

14. The fast-slow guidelines are from forestry sources, i.e., Schenck (1904), Schlich (1910), and Yoshida and Kamitani (1997), modified for general use by this author.

15. Timing, the planting of the shade tolerant, fast grower first, is another option. This would have greater consequence if the fast grower is deep rooted, the light demander is shallow rooted. This opens another road, one not explored here, that of finding winning species pairings based on matching compatible above and belowground attributes.

16. Row orientation and vertical–horizontal light is a bit more complex than this. Row orientation, as a design tool, is further explained in Wojtkowski (2006a, p. 172) or Wojtkowski (2002, p. 198).

17. As entities unto themselves, there are indices to gauge agrodiversity (Coffey, 2002). As these do not relate to any economic measure, they are not used here.

18. The number of applied examples of PPC is very small. Ranganathan *et al.* (1991) calculated the PPC for a leucaena–pigeonpea facilitative system (this resulted in a below one LER) and for a pigeonpea–sorghum intercrop (the LER was greater than one). The PPC has and continues to be advocated and used for abstract explanations (e.g., García-Barrios and Ong, 2004).

19. There can be disagreement on or climatic situations where there is no one single optimal monocultural planting density.

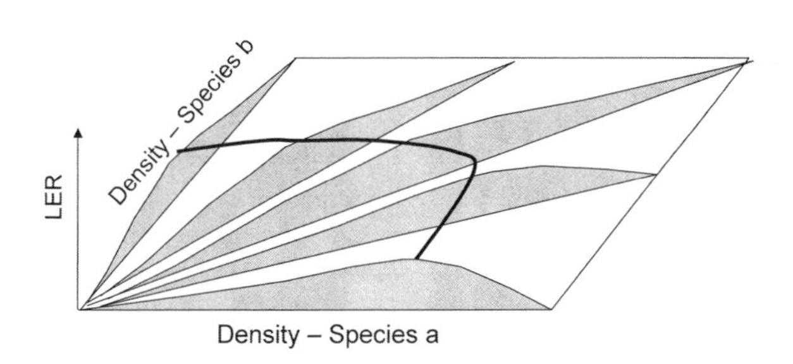

FIGURE 1*a*

20. Note that Figure 3.2 underlays and is an overview of Figure 4.4.

21. Figure 4.4, with its third dimension, contrasts to more traditional PPCs as presented in Figure 4.5. Under the traditional two-species PPC, the furthest point from the origin equal to the topmost value (Figure 4.4). In a simplification below, the 3D PPC is derived connecting the optimal points on the individual ratio lines, not as a continual surface (as in Figure 1*a*). This is closer to more traditional method of PPC calculation.

22. Rather than an RVT-based PPC, the traditional means for adding price has been to use a PPC, but with a price tangent. This method, not repeated here, is found in many agricultural economics texts.

23. The maize–soybean LERs are from Prasad and Brook (2004). The ending LER was between 1.3 and 1.45.

24. Each basic pattern, of which there are many, offer numerous variations or pattern arrangements. For a row pattern with two species, a and b are represented in cross-section, the variations include, abababa, aabaabaa, abbabba, and aabbaabbaa. Only touched upon here, spatial theory is valid topic with extensive ramifications (Wojtkowski, 2002, p. 50).

25. The late planting of beets with wheat is from first person notes (Anon., 1872). Also included are shallow-rooted barley, rye, or clover with deep-rooted onions, peas, or cabbage.

26. The proof for this, where ecosystems are able to resist the forces of nature (e.g., fire, drought, etc.), is found in numerous natural examples. Although resistant in their established form, thriving ecosystem, once removed from a site may, because of adverse site conditions, fail to reestablish.

27. This statement on the non-quantifyability of complex agroecosystem is based on the lack of data from agroforests. The other example, pastures of mixed grasses, is far easier to measure (as in Tilman *et al.*, 2006).

28. Scientific classification does not involve product differentiation. As examples, varieties of *Brassica oleracea* produce some very different outputs (cabbage, broccoli, cauliflower, or kohlrabi) whereas the outputs of the genera of *Triticum* spp. (wheat) can be hard to distinguish.

29. Multi-varietial interplanting for rice is found in Yoon (2000), for wheat, MacDonald (1998) and Plucknett and Smith (1986), and maize, Altieri and Trujillo (1987). Slicher van Bath (1963, p. 262) lists old European grain mixes (e.g., wheat with rye, spelt with rye, wheat with barley, rye with oats, etc.). The reason for these mixes is to lessen risk.

30. This refers to the production of alcohol (ethanol) as a motor fuel, e.g., from sugar as produced in Brazil.

31. Details on the maize-bean intercrop are in Davis *et al.* (1987) and Davis and Garcia (1983) with minimum rainfall levels (Rao, 1986).

32. Breaking this common prescript, there are simple intercrop mixes that maximize the use of site resources, doing so with cost efficiency, e.g., through perennial species. If cellulose is the base stock, the production of ethanol can go beyond a varietal interplanting (as mentioned in an above endnote) and become a simple mix. But why stop with a simple species mix, adding trees could further improve the ecological outlook.

33. As a topic, buffer species are discussed in Chapter 13, section 'Buffer Species.'

34. Buckwheat around cucumber is from Platt *et al.* (1999).

35. The magnitude of the livestock gains is in Prinsley (1992). Although for windbreaks, this does apply to the shelter afforded by a single large tree.

36. There is a published study of tea strips alternating with strips of forest trees (Wang, 1994). Better micro-climate and frost protection are the reasons cited.

37. The German streuobst is from Herzog and Oetmann (2000).

38. The cucumber with orange intercrop was observed in Central Mexico by this author.

39. The light shade examples are from Ashley (1986) and Anon. (2003).

40. Some counts of contained-agroforest plant and bird species are in Cooper *et al.* (1996).

41. Discussions of agroforest variations continue in Chapter 14, section 'Complex Agroecosystems.'

5 Biodiversity

In nature, many plants play a facilitative role, either helping other species or making contributions to the ecosystem in which they reside. In an agricultural setting, facilitative species do likewise. It is these plants that open other economic possibilities. Instead of agroecosystems driven by multiple outputs and by an economic emphasis on increasing revenue, facilitative species, by replacing inputs, cut costs.

Cost-reduction systems are not always the providence of resource-poor subsistence farmers, but can also be economic strategy for large commercials farms. Within the realm of cost orientation also comes strong theoretical underpinnings, both agrobiodynamic and economic. Often sidestepped in the literature, this track, and the associated agrotechnologies, may divine the future of agroecological economics.

APPLICATIONS

Biodiversity, in economic terms, is the science of replacement. Instead of confronting shortfalls in essential resources, large populations of plant-eating insects, or plant-attacking disease through ex-farm inputs, the user attempts to utilize biodiversity as nature intended. In doing so, a slew of potential problems, economic and/ or environmental, can be addressed. It goes without saying that, in a mixture of productive and non-productive species, those that produce are the primary species.

UNDERLYING BIODYNAMICS

The biodynamic principles offer plenty of avenues for guiding biodiversity. Facilitative plants, those that aid the primary species, can be added. These secondary species generally offer little in the way of direct economic benefit, but plenty of secondary gains.

There is a long list of direct plant-on-plant services that can substitute for outside, at times expensive, inputs. The expected gains are much the same as with agrobiodiversity, only the effect and outcome, because plants are totally dedicated to their task, should be a lot stronger. The possible gains lie in

- indirectly increasing the plot land equivalent ratio (LER);
- providing protection against plant-consuming insects;
- keeping weeds at bay;
- offering a more favorable micro-climate (e.g., greater humidity, less plant drying);
- moderating yield-robbing temperate extremes;
- enhancing water inflation and reduce surface runoff;
- buffering unsuitably toxic soils (i.e., to acid, alkaline, salty, or to loaded with allelopathic compounds).[1]

Yields

From Chapter 4, interception and/ or conversion are the chief mechanisms of gain when all component species are productive. There are limits, two separate species, acquiring and dividing the same resources, cannot have an LER that exceeds 2.0. This only happens when two closely planted, dissimilar species co-exist in relative resource abundance or they draw their limiting resources from different sources.

This is where facilitation becomes important. The biculture upper limit of 2.0 can be exceeded, at times, by a good margin, e.g., values of 3.0 or higher. This happens when the gains of productive species (one or more) are amplified through plant-on-plant facilitative affects.

With three productive species, the acquisition and partitioning LER will most likely be lower than 3.0. The ecological logic holds that there are not enough niches to go around. Again facilitation, by proving what is missing, comes to the rescue, further driving up the LER. This will continue with four, five, or more species (four-plus polycultures). If the LER is to exceed the number of productive species, strong facilitation must be present.

Adding more species helps. However, as more are added, this presses against two unknown, but real, ceilings: (1) those of acquisition and/ or partitioning and another, (2) when facilitation is factored in.

LER and related ratios estimate resource-use efficiency, not overall productivity. Poorly yielding sites can have high LERs if the few resources available are used with effectiveness. Ratios indicate if the direction taken, and the facilitative mechanisms being utilized, are helping. In this case, facilitation can enrich poor sites, increasing both the LER and the actual value of the harvest.

Cost Reductions

The often overriding reason for biodiversity is cost reduction. The notion being that outside or ex-farm inputs can be reduced or eliminated if a second species assumes the required tasks. This can be measured comparatively through the cost equivalent ratio (CER) or through actual cost figures.

Substitutable Inputs

The reliance of conventional commercial agriculture outside, ex-farm inputs does promote higher yields but, with frequency, at an environmental cost. The costs referred here to are not always directly economic, some are intangible, e.g., the health of nearby natural ecosystem or the effect on clean water. Biodiversity substitution, e.g., for environmentally harsh insecticides, will accomplish yield goals with a less destructive input.[2]

Substitutable inputs include:

- fertilizers,
- insecticides,
- herbicides,
- fungicides,
- manual weeding,
- irrigation.

Desirable Plant Characteristics

Whereas the success in pairing multiple productive species relies more upon complementarity, the selection of non-productive companion species relies more on desirable plant characteristics (DPCs). This is because there are far more companion choices increasing the opportunity for very few salutary pairings. Whatever the need or needs, a facilitative plant can be enlisted to provide it. For enhancing the productive potential of the primary species, examples abound.[3]

One need is weed suppression. This comes in varying forms. Some farmers selectively weed, removing only those weeds that effect the crop, leaving those that do not. Those that remain establish, without planting, an effective weed excluding covercrop. More often, this is done through purposely planted cover species.

A lot of species are available and, through DPCs, do serve well. Generally, ground-hugging plants with big leaves are preferred.

Examples are plants of the squash family. Other options are species with small leaves which are dense upon the ground. Small-leaf, weed smothering plants include crown vetch. As a secondary species, this has been intermixed with sorghum or cotton. Other small-leaf cover-crops are rye grass or white clover. These have found use with pepper or okra.[4]

The mix of DPCs is all important. If may be better to find a species that has large leaves while, at the same time, fixes nitrogen, repels insects, controls erosion, and does other ecological tasks. Multi-tasking may seem a tall order, but with so many plant species from which to choose, one need not always settle for a lesser choice.

Cut-and-Carry

Rather than direct plant-on-plant interaction, the better option may be a nearby location, one with enough distance so the plants do not negatively interact. These are the cut-and-carry design where the facilitative species is in a neighboring block or strip.

Plants that repel insects can be so planted, at the ready when an outbreak threatens. Another substitute, in this case for herbicides, is weed suppressing mulch.[5]

There are many advantages to a cut-and-carry strategy. Among these are that:

- it is possible to pick facilitatory species (one or more) for their nutrient gathering or other properties, not their plant-on-plant competitiveness;
- the cut-and-carry schedule can be formulated around crop nutrient requirements, with less need to focus on plant-on-crop light, nutrient, and water competition;
- the cutting of biomass can be spread away from peak use periods with increased overall labor efficiency;
- there exists the opportunity to direct the biomass to other plots and other uses (e.g., as animal feed);
- the biomass strip can be laced with insect attracting, insect toxic, or insect repellent species, as part of an insect control strategy, directing insects away from the crop plots.

Among the disadvantages are some inefficiencies in resource capture and transfer. Root biomass from the facilitatory plant is not part of the nutrient transfer mechanism (those cut on-site through plowing and allowed to decay). Another disadvantage is the cost of the carry portion of a cut-and-carry strategy. Despite these drawbacks, plots of facilitative species do find use in increasing the LER of nearby productive crops.

Risk Abatement

Interactions occurring with secondary species can minimize potential crop losses. Facilitation, besides boosting yields, helps when essential resource shortfalls occur or when rainfall and/ or winds can be overwhelming.

Rain Impact

Not often noticed, but of importance, is the force when raindrops impact the ground. This can dislodge small stones and particles of soil, making them more susceptible to being carried away by flowing water.

Rainfall impact can be countered by directly overlaying the land with living vegetation, biomass, stones, or some other protective covering. It should be noted that covercrops are particularly effective because these are close to the ground. In contrast, trees of great, or even intermediate height, are less effectual because drops of water, falling off leaves, do strike the soil with much the same force as an unabated rainfall.

As complete protection, a good cover species will absorb the impact, hold in place the soil and the water, and promote infiltration. Also required is year-round protection. If a ground cover is not always in place or if in place can be overwhelmed, then other defenses may be required (see mounds, Chapter 8).

Climate

If the regional climate is hostile to good yields, the micro-climate within plots can be of paramount importance. Masses of vegetation, especially overstory plants, can trap moisture and moderate in-plot temperatures. Any reduction in transpiration caused by a lessening in wind flow and evaporation can extend scare water, making a low yielding plot into something better.

In climate altering designs, some types of biodiversity do not change the shape of the production function, these only shift the relative positioning of this function via the ambient temperate, a limiting essential resource, or some other resource. The resulting critical shift provides a conceptional view of what will happen when landusers make modifications, vegetative or otherwise, to coax yields on reluctant sites and in hostile climates.

Figure 5.1 shows how the production function can change in the face of temperate moderating ecosystem. The critical shift fits a number of different scenarios, beside intercropping with a facilitative species, this applies to rotations, landscape factors, land modifications; all of which can shift change the production function to better fit a climate, i.e., rainfall levels, rainfall patterns, ambient and/ or extremes in temperatures, etc.

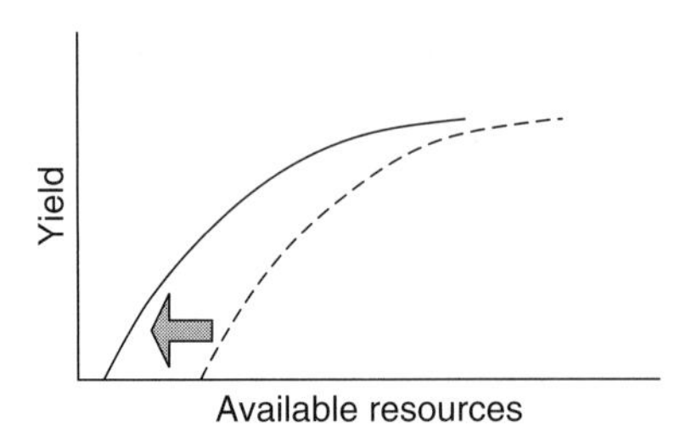

FIGURE 5.1. The critical shift, what happen when a plot environment is altered. The net effect can be to shift leftward a species production function with regard to the resource or resources in question.

Insects and Diseases

Many insect control methods fall under the heading of facilitative biodiversity. The plants involved provide habitat for beneficial predator insects or serve to repel, decoy, or lure away unwanted herbivore insects. The economic gains come in reducing or eliminating the need for ex-farm chemicals.

Given a favorable habitat, predator insects can be welcomed. Spiders, ants, and a very wide range of insect-eating insects can be part of a biodiversity strategy. The idea is to keep the populations of detrimental insects at base, non-threatening levels by insuring a favorable local (in-field) habitat for the various predator types. Absent insects upon which to dine, the predator populations will fall to the point where these insects may become ineffective as a deterrent. It is therefore presumed that, in encouraging at least a few plant-eating insect types, that some crop losses are expected. The level, or value of these losses, must be less than cost of other, more immediate controls, e.g., continued chemical sprays.

Not all countermeasures require crop losses. Selected plant species have the ability to repellant specific types or a range of insect species. Added to an agroecosystem, these plants can serve a specific role. Since no economically interesting outputs are expected, the cutting or damaging of the leaves to release volatile, leaf-contained, insect-averting chemicals into the air is a viable option. Although labor intensive, this practice is helpful to contain outbreaks of harmful insects.

There is wide range of insect repellent plants. Placed next to susceptible crops, these have proven, time again, to be effective. This is not always an aboveground mechanism, another use is to counter in-soil nematodes. As one instance, nematode-resistant varieties of cowpea protected tomatoes.[6]

Less utilized, and with some attending danger, are plants to lure insects away from the crops. The danger comes in that insect can multiple on the decoy plants, later re-infecting the desirable species. As such, decoy plants are only utilized in conjunction with some other strategy, e.g., decoy crops can be a lure when planted near those plants that harbor predator insects.

As for plant diseases, micro-climate slows the spread as does transmission-interfering biomass. In this, less is known and the possibilities are less developed, including predator diseases that inflect and weaken other diseases. Despite what may ultimately turn out to be a vast array of disease-countering tools, most of the biodiversity effort centers upon blocking plant–plant transmission, i.e., containing the spread of a malady.

Relevant Guidelines

In placing a productive and economically valuable species near one providing only facilitative services, the key is having the facilitative plants provide more of the limiting resource to the primary species. A nitrogen-fixing species next to a primary species that demands this element is the frequently cited example.

A system may be more successful if the facilitative species offers more, in terms of facilitative services, than the proffering of a single mineral resource. With a multitude of non-productive facilitative species from which to choose, tasks, such as insect or extreme temperate control, can also be addressed through a single facilitative species. This means that the DPCs for a purely facilitative species can be more accommodating than possible through agrobiodiversity.

The result is one of the key intercropping guidelines (as restated from Chapter 3). Because of greater pairing possibilities and expanded potential gains, facilitative biodiversity is best employed on resources poor, less agriculturally attractive sites. In contrast, agrobiodiversity, with far fewer interplant facilitative services and with the opportunity for marginal gains (higher LERs), is more appropriate on more fertile, well-watered sites.

Beyond this, there are other biodiversity guidelines. Above the ground, a facilitative species is almost always the understory except when the facilitative affect involves blocking excessive light. The result, facilitative species are either shorter or much taller than the primary crop. Canopies of equal height would be an unusual case.

Weed control is an understory effect, the facilitative species is often shorter than the primary crop as occurs with covercrops. This may not always be the case, a weed control facilitative species may rise above the primary species. There is logic in this. With one canopy, weeds can survive, with two light-intercepting canopies, the weeds are at a severe disadvantage. In this case, the crop, residing below the facilitative species, should be shade resistant.

ECONOMIC MEASURES

In contrast to omni-productive systems, facilitation requires a different economic handling. There are systems where high yields are paramount, others seek cost savings. The emphasis as expressed through biodiversity is often one of cost-orientation with the accompanying economic measures.

A large portion of facilitative biodiversity seeks to replace inputs or methods being utilized with something more cost effective. These substitutions are gauged byway of the crude CER (equation 2.4) or the price-adjusted CER (equation 2.5).

Adjusted CER Demonstrated

Case studies on CER use are far from abundant. However, the literature does contain an example of economic orientation in monocultural coffee and shaded polycultural coffee (Perfecto *et al.*, 1996). The data for this single output system have Y_a as the yield of monocultural coffee; Y_{ab} the yields with a non-productive shade tree; C_{ab} the costs with shade trees; and C_a costs without such trees. From this,

$$\text{LER} = \frac{Y_{ab}}{Y_a} = \frac{\$314}{\$1397} = 0.22$$

$$\text{CER} = \frac{C_a}{C_{ab}} = \frac{\$1740}{\$269} = 6.46$$

and

$$\text{CER}_{(\text{LER})} = (\text{CER})(\text{LER}) = (6.46)(0.22) = 1.42$$

In this case, the LER is very low (0.22), but is not indicative of economic potential.[7] Clearly, monocultural, high-input coffee out yields the polycultural system. What is attractive are the very low costs of production. The crude CER equation, specifically, \$1740/ \$269, shows shaded coffee is almost 6.5 times more efficient in the per area use of inputs.[8]

There are two analytical tests here: (1) the very high CER value (6.5) indicates that scarce external resources are being utilized efficiency; and (2) a $\text{CER}_{(\text{LER})}$ value greater than 1 (1.42) shows the reductions in yields more than compensated for by the cost gains. In this intuitive, concisely expressed form, the result demonstrates why shaded coffee is favored by resource-poor farmers.

Cost Floors

There are questions on how far down the cost cutting road a farmer can go. With revenue orientation, one never adds inputs when the potential revenue gain is less than the cost of the inputs. It goes without saying that, for cost orientation, substitution are never made when the loss of revenue is greater than the savings obtainable.

There is another cost aspect. In a complication not found with revenue orientation, cost orientation has a sowing/ harvest floor. An agroecosystem will never be profitable if the revenue received is less than the cost of land preparation, sowing, and harvesting. This puts an effective lower limit on how much cost-cutting and revenue loss can occur.

As expected, economic strategies have been devised to lower this floor. Through perennial grasses, the common pasture eliminates sowing costs.[9] Going further, some utilize semi-husbandry, animals raised in a semi-wild state with little or no maintenance. This concept is expanded in Chapter 14.

The cropping equivalent can be a gathering strategy in uncultivated areas. Wild rice is one crop that can be seasonally harvested from naturally seeded wetlands. Some regard the ratoon of planted rice (the regrowth after the first harvest) as a lowering of the cost floor. Other forsakes the low yielding ratoon as an uneconomic cost-oriented land use, preferring the revenue orientation of a newly planted crop.

Cost Possibilities Curve

Where facilitation is directed toward boosting yields, there is a requirement, as with agrobiodiversity, for the best spatial pattern and densities. The production possibilities curve (PPC) is a tool to optimize these variables. It is equally helpful to derive a cost curve similar to that of the PPC. As in Figure 4.4, there are ratio lines with associated costs. Connecting those points that represent the lowest costs, using crude CER values, delivers a CER curve (as in Figure 5.2).[10]

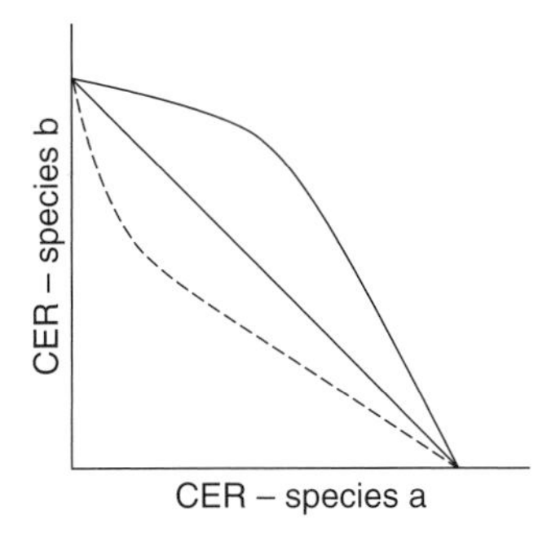

FIGURE 5.2. The CER curve with the outermost point being the most cost effective for the intercrop in question. The inverse of this function (the dotted curve) may be more intuitive.

DESIGN VARIABLES

As with agrobiodiversity, species pairings, temporal adjustments, planting densities, and spatial patterns, adding to the array of possibilities, can confuse the topic. The agrobionomic principles help in sorting through the possibilities when there is scant empirical evidence upon when to formulate a design.

Specific Interaction Zones

One aspect of spatial pattern is the specific interaction zones (SIZs). These determine the interspecies distance. For some uses, it is not necessary to blanket the plot with the facilitative species. If the zone of influence is known, this will determine spacing. If an insect repellant plant drives away insects for a 2-meter distance, then a 3–4 meters between-plant spacing captures this influence without undue interplant competition and/ or without uncalled-for planting costs.

Spatial Patterns

Where co-inhabiting productive species are highly complementary, high-interface, fine patterns (as in Figure 4.2) and high per area planting density can be a sure route to a high LER. This is also a strategy when pairing a single productive species with a single facilitative plant. The assumption being that the one facilitative plant is fully capable of handling the limiting resource needs of a paired productive species.

This is not always the case. In order to make up a deficiency in-site essential resources, one productive species may need to be accompanied by a number of facilitative plants (as in cross-sectional Figure 5.3). Two of the many possible spatial patterns for this are shown in overview in Figure 5.4.

As with the agrobiodiversity approach, many of the facilitative agrotechnologies carry, in their formulation, an inherent spatial pattern. It is contingent with the user to expand upon this, finding the best planting densities to bring a system to an economic best.

Timing

In some roles, timing is all important. This happens with weed control in seasonal cropping. Planting the main crop early or late can help regulate weeds. Pertinent questions also arise with covercrops. When planting anew each season, there is the question of one or two plantings. It may be cheaper to simultaneously plant both the main crop

FIGURE 5.3. In cross-section, an uneven ratio, two ground-hugging facilitative plants for each productive species.

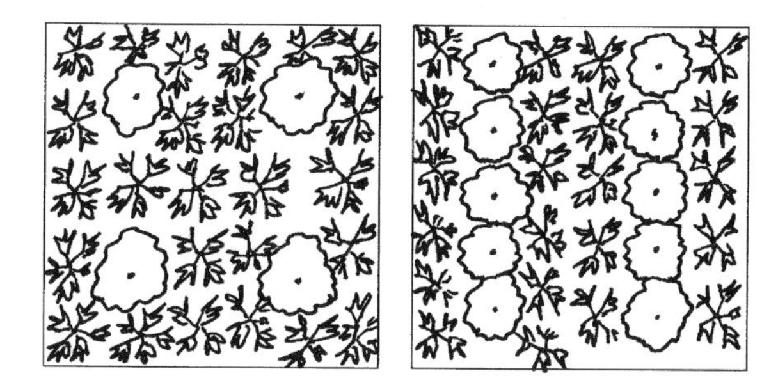

FIGURE 5.4. In overview, two different planning arrangements for the pairing of many facilitative plants with a few productive species.

and the covercrop, however, a temporal separation, allowing the main crop time to establish, may reduce interplant competition.

Not only a weed issue, timing may also effect insect dynamics and, when rainy seasons are short, the efficiency of water use. Timing is also an issue when applying nutrient-carrying biomass.[11] In general, the dynamics of timing, although all encompassing, is a relatively unexplored topic.

Pruning

Treecrop or tree–tree association can require management inputs to alter the internal plot dynamics, i.e., the mix of essential resources going to each component species. For woody plants, this means some type of pruning. Of the techniques, there are (a) coppicing, (b) pollarding, (c) stem pruning, (d) lopping, and (e) branch pruning.

Coppicing cuts woody plants to the ground level. The result should be many stems with bushy, low canopy. This is ideal for the production of leafy forage, green manure, or mulch.

Pollarding is similar to copping, except the stem cut high, usually about 2 meters. aboveground level. Again, this produces a multi-stem, bushy canopy that is above and out of reach of grazing animals. Reasons are the same, forage, mulch, green manure, the difference is the area can be grazed without losing the tree canopy to hungry animals.

Stem pruning is classic forestry pruning technique. The lower branches are cut close to the stem. The purpose is a clear, branch-free stem of high worth as defect-free lumber. As a multi-species accommodation technique, the reason can be to move the canopy higher, evening out the light that reaches the understory crop.

Lopping is the process where the outer branches of a canopy are removed. More costly than the other techniques, the result is a smaller-diameter, internally denser canopy. This is useful when direct light over crops is desired, i.e., when crops are placed in the intratree space, as is the case with treerow alley cropping.

Branch thinning is where the main branches are kept and the secondary branches are removed. This opens the canopy for greater light penetration. This is useful when higher levels of filtered light, rather that low light levels or large doses of direct light, favor the understory crops.

Weeds

How weeds are naturally controlled is first and foremost determined by relationships of weeds and the primary species. If the primary species has good suppression properties, weeding is done away from the stems, mostly in the interrow space. If this is true, a weed suppressant covercrop should be planted in and occupy the between row space. As always, the covercrop must be complementary with the primary species. This situation offers flexibility in the DPCs of the cover species.

If the primary species is effected by and has poor weeds control properties, one might seek covercrop that thrives better beneath the primary species. For this, a higher degree of interplant complementarity, as a DPC, is required.

In general, weed control is not an easy topic. The vigor of plants may be more important than the intercrop combination. There are also niche questions, some intercrops suppress some weed species, but do not work well against others.[12]

ASSOCIATED AGROTECHNOLOGIES

The defining characteristic of biodiversity, in all its forms, is that nonproductive plants support, often through increased yields or reduced costs, plants that do provide an economically useful output. A number of agrotechnologies have been formulated with this in mind. Keeping in mind the degree of overlap these have with agrobiodiversity, the listing of biodiversity agrotechnologies is:

- Perceived monocultures
- Facilitative intercropping
  - Simple mixes
  - Strip cropping
  - Boundary
  - Covercrops
- Facilitative agroforestry
  - Parkland
  - Protective barrier
  - Alley cropping (hedgerow)
  - Strip cropping (mixed tree)
  - Crop over tree
  - Physical support systems
  - Shade systems (heavy)

Perceived Monocultures

It may seem strange to find a monoculture under the biodiversity heading, however, this is a common variation. Farmers often view and manage systems as a monoculture ignoring large populations of other species (Photo 5.1).

PHOTO 5.1. A perceived monoculture, in this case an apple orchard with an ecologically disregarded understory.

An example is an orchard with a ground cover. There is little visual difference between perceived monocultures and those with plenty of unwanted weeds. Perceived monocultures are slightly more cost oriented as farmers trade expensive weeding for cheaper understory mowing. In doing so, they might experience a yield penalty. The economic issues lies in weeding costs vs. mowing as well as the land cover gains (e.g., better erosion prevention) vs. the yield penalty in tolerating a non-specific ground cover.

These could be much more cost oriented if full advantage is taken of the biodiverse understory, e.g., incorporating insect control ground cover as part of the species mix. In taking this additional step, this is no longer a perceived monoculture, but rather an ecologically functioning covercrop ecosystem.

Facilitative Intercropping

There are a number of applications where a seasonal facilitative species will find use. To be economic, simultaneous planting is often a prerequisite. More often, it is better if the facilitative species is a perennial as this eliminates the cost of one planting, bringing about stronger cost-orientation.

Simple Mixes

The simple mix has facilitative species interplanted with productive species. Nitrogen-fixing cowpea with nitrogen-demanding maize, where the cowpea is only present as a facilitative addition, is one such case.

What makes this a simple mix, rather than a covercrop, is the relative height of facilitative and the primary species. The non-productive addition can be taller than or equal in height. In an already mentioned example, cabbage, in seeking vertical light, can easily push aside taller grasses. The issue is what happens within, and how competitive is, the belowground situation.

If a facilitative species is to provide maximum worth, close plant-on-plant contact (interface) and find patterns (as in Figure 4.2) will better confer whatever benefits are expected, e.g., predator insects are closer a hand and presumably more effective or, also through a dense ecosystem, where the internal micro-climate is less changing.

The second species can, by adding what is missing, increase costs and yields (revenue orientation), through substitution for expensive inputs, lower costs with a yield penalty (cost orientation), or simply reduce the cropping risk. As plant-on-plant governed systems, much depends on the DPCs of the second species. Once a good match is found, the problem evolves to one of densities, patterns and, in some cases, timing.

Strip Cropping

The facilitative version of strip cropping can be independent from or a rotational adjunct to the agrobiodiversity version. In alternating strips of productive and non-productive species, nutrient gains are often coupled with some form of insect control, e.g., repellant plant. The non-productive biomass can be hand cut-and-carried or machine mowed-and-tossed onto the productive strip.

The advantages, other than these being farm machine friendly, are enumerated in this chapter (under cut-and-carry systems). The economic answers lie in the vast array of species and spatial possibilities as well as the timing of the cut and carry. Some facilitative strips can be revenue oriented, others cost oriented, depending on whether the strip biomass is directed toward higher yields or economizes on imported nutrients.

Boundary

Common in some regions, farmer ring or subdivide crop plots with a non-woody protective species. As in Chapter 4, these can be productive plants, with or without the non-harvest option. The boundary can be a single row or in multiple rows. The active mechanism is often the windbreak effect although other ecological tasks, e.g., controlling the spread of damaging bugs, above and/ or below ground, may overshadow the windbreak gains.

A bit more is expected from a perennial barrier. A dense curtain of vertical roots will impede the root spread of the juxtaposed crops, allowing plots or strips of highly competitive crops to be closer that otherwise possible.

A non-productive, non-woody species, one with the desired DPCs, is the tropical grass vetiver. This species is very deep-rooted, is maintenance free, perennial, with a tall vertical canopy. It has the added advantage in that it will not spread to become a weed.

The basic economic decision, to use or not, is comparatively simple. The question is, do the yield losses, through the space taken by surrounding or subdividing plots with a boundary species, result in better yields on nearby lands (as is the case in Figure 15.2). The second phase in the economic analysis is the optimal pattern and plot sizes.

Covercrops

Covercrops represent a special case in simple intercropping. The principle task is weed control. To avoid inordinate interference with the primary species, the facilitative plant always shorter than the productive species (one or more). This understory can be seasonal or perennial.

In addition to weed control cost reductions, it is not too much to expect yield gains by way of nutrient or other in-soil facilitation or through favorable insect dynamics. There is ample evidence to support these results.[13]

As a popular agrotechnology, the list of suggested cover species is long, the attributes (DPCs) of each complex. Because of this, the yes or no question, to use or not, is often lost in the details.

If a covercrop is directed, by way of facilitation, to higher yields, the result is revenue orientation. Often, what is expected is input substitution, the cost of the covercrop being lower than manual weeding costs and cost orientation. The same happens when yield reductions are balanced against cutbacks in herbicides and hopefully insecticides.

Tillage, the longevity of the covercrop, the gains, cost or revenue, conferred, and the complementarity of the proposed cover species all enter in the economic picture. Longer lived, naturally reestablishing, or perennial cover species might have an ecological and economic edge over short-duration species. As all the options have not been fully explored, this is not an easy analysis.

Facilitative Agroforestry

The most cost-oriented facilitative systems are those in agroforestry. This is because trees are long-lived, planting costs can be amortized across many seasons, and tree are less costly to maintain. Additionally, trees often survive adversity better than annual species and, when maintenance is required, this can be done off-season when additional labor is available. The disadvantage is that the trees are fixed in-place and expensive to remove. If the current market for a treecrop is weak, switching tree species can be problematic.

Parkland

The parkland system has productively inert trees widely scattered across spacious fields. The trees can be ordered or unevenly spaced over crops or pasture. The revealing characteristic, other than the wide intratree distances, is the use of one or two designated parkland tree species in a region.

Protections afforded by the trees include stopping wind microbursts, harboring birds, increasing water infiltration. The intent is to duplicate the dynamics of natural savannas for the benefit of crops or a grass with animal component. For cropping, the trees are predominantly deep-rooted, for the pasture version, the trees tend to have evolved for a grazing environment, e.g., tree species often characterized by large thorns.

PHOTO 5.2. A pasture parkland of classic layout and composition, i.e., scattered trees of a single species. This photo was taken in Nicaragua.

These systems also address a food security issue. This can be important in some regions. If the crop fails through drought or insect plague, the trees offer an insurance harvest. In ancient Europe, the acorn served this purpose, being available when other harvests were lost. An existing example is the dehesa system of Spain and Portugal. This has a mix of oak trees and grasslands, both of which support hogs and cattle.[14]

The economic decision is mostly yes or no, i.e., do the benefits outweigh the disadvantages. There are upgrading variables. Tree pruning can reduce tree-on-crop competition. The pruning decision is made easier if the pollarded tree branches have an end use or a market (e.g., as firewood). Parkland trees in a cost-oriented pasture setting are seldom trimmed.

There might be enhanced ecological returns if parklands are one part of the larger landscape. Wind reductions can be based on widely spaced windbreaks and parklands (see Figure 12.4). Another assenting use is as horqueta trees. For this, platforms are placed in the trees upon which forage is laid out of reach of animals. Parklands of horqueta trees are found in the sub-Sahara and in drier parts of South America.

Add-ons in the form of agricultural bat or birdhouses deepen the system with regard to insect control or, if owls or hawks are involved, mice and rat control (see birds, Chapter 11). In this environmental setting, the trees function as bird roosts from which to hunt. Additional products (e.g., firewood harvests) and a few intangibles (e.g., aesthetics) can swing the economics in a favorable direction.

Protective Barrier

As with seasonal barriers, the agroforestry (tree or shrub-based versions) confer many of the same advantages, i.e., the contain the spread of insects and the like. In offering a tall, dense barrier, especially at ground level, these also have windbreak and other possibilities.[15]

One unique use of the tall, dense perennial barriers is in containing the drift of chemical inputs. Insecticides, herbicides, fungicides, etc., are more efficient if localized. In this way, they accomplish their purpose without interfering with neighboring natural and agricultural ecosystems.

The cost, in terms of lost yields and the inputs required, should be negligible in comparison to the gains. Barriers should also prevent roots spreading from one plot to another[16]. As a counter against a particularly pesky insect or plant disease, barriers often prove advantageous.

The economics are much the same as with non-woody boundaries but, since these can be taller, their influence extends over a greater area. To the negative, such trees are fixed in place and costly to remove. As a result, many farmers place these in less intrusive locations, e.g., near farm boundaries, roads, steams, and other fixed farm feature. If plots require further protective subdivision, a seasonal or an easy-to-remove perennial is employed.

Alley Cropping (Hedgerow)

Hedgerow alley cropping is the growing of crops between a row of pruned woody hedges between crop strips. The hedges are usually about 1-meter tall.

Maize is a common crop, but crops are found in conjunction with a suitable, nutrient-providing hedge species. The system, hedge-on-crop gains are facilitative. This is dramatically expressed in the fact that maize alley cropping produced some of the highest LERs recorded (e.g., 3.69). This points out the potential, especially on nutrient-poor or overworked soils.[17]

This number alone can paint an overly rosy picture. The system has many limits, tree-on-crop competition for moisture being notable and questions as to which hedge species are best with crops. The main requirement seems to be a good to excellent DPC match with the crop or crops. Good hedge-crop matches have found, but through much trial and error. In a tree-tea system, five of the tested tree species reduced yields 22–40%, the sixth tested increased outputs 23%.[18]

There are other economic aspects. Hedges are costly to prune, permanent and expensive to remove so cropping flexibility is not assured. This is especially true if the hedge species turns out not to be compatible with one or more crops. This may limit the crop rotation possibilities.[19]

The management sequence enters the picture as the hedge must be trimmed and the biomass nutrients released in time to meet the demands of the crop. The economic questions, beside an immediate harvest improvement, are with yield sustainability. Are the hedges, as the primary vector, sufficient to insure continued cropping without rotations or other soil-boosting measures?

With all these concerns and possibilities for improvement, this becomes more than a simple yield–cost analysis. It helps greatly if the branch prunings from the trees have a marketable end use, most often as firewood.[20]

Mini-hedges are a variation of the hedge design.[21] Instead of being a convenient height for hand-pruning, these are cut slightly above ground level. This variation is intended for mechanized farms where the cutting is done with a tractor-attached brush cutter. The economics are a bit different, no firewood is expected. Hedge pruning is measured against yield increases and the effect of cut biomass in reducing weeds, soil erosion, and increasing moisture retention.

Tall or short, the substitution is away from fertilizers, which a landuser cannot afford, to labor, which the user has. If the motive is greater productivity or revenue through more labor inputs, alley cropping is a revenue-oriented activity.

Once some basic parameters are known, this agrotechnology may be a far better on erosion-prone hillsides. Although there are a few intangibles associated with alley cropping, the propelling motive is increased yields. Having a double purpose (yields plus erosion control) may make a hedge system economically attractive.

Generally, a well-designed system can be an economic plus. Poorly formulated, these are an economic non-starter. With so many variables and the need for a good treecrop pairing, positive tree-and-crop pairing are hard to assess. In rejecting hedgerow alley cropping, there should always be a nagging feeling that, with a little more development, the decision would have been quite different.

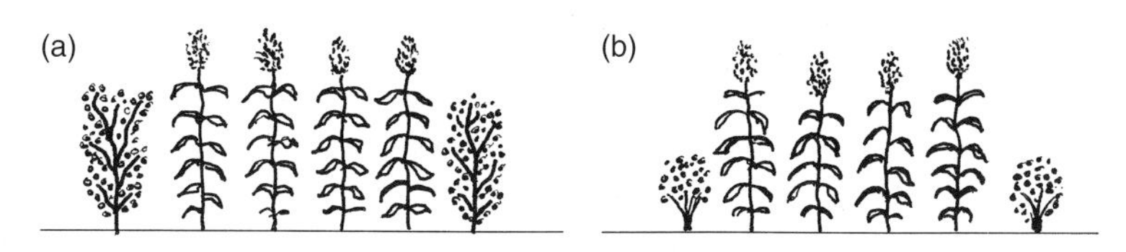

FIGURE 5.5. A cross-section of two hedgerow formulations: (a) normal hedges and (b) a mini-hedge.

Strip Cropping (Mixed Tree)

As a facilitative version of an agrobiodiversity system, mixed tree, strip cropping utilizing wide strips of trees, rather than one or two rows, are comparatively rare. The main purpose is to transfer nutrient-rich biomass directly to the crop strip. Where a single row hedgerow may not provide enough biomass, a wider strip can.

One drawback is the flexibility issue: once planted, tree strips, being expensive to remove, tend to remain. What is a shortcoming to some may be a blessing to others. Strips of pruned woody plants can be less costly to maintain and less subject to degrade under the extremes of a climate.

As with the hedgerow alley cropping, the trees can be taller for hand pruning or short, i.e., a mini-hedge, for machine mowing. If conditions are poor for the crops, the biomass can ensure grazing. With this version, the woody strip is crop facilitative during good years (by providing nutrient-rich biomass) and a backup forage source when the main pastures wane and the crop strips fail to yield. This dual purpose, spanning both good and bad years, makes this a risk-attractive agrotechnology.

The economic question is whether the land area removed from crop production and put into biomass production results in a net economic gain. In addition to risk reductions, the cost–benefit equation includes the yield gains brought about by the nutrients transferred, the mulch-induced weed reductions, the strips as an anti-insect countermeasure, and through reductions in wind and water erosion. As with alley cropping, there are few, if any, intangibles to be considered.

Crop Over Tree

There are situations where the trees can be shorter than the crops. Planted in a fine pattern (see Figure 4.2), this allows for a close interface between the tree and crop. Competition is reduced through continued pruning (i.e., coppicing) of the stems. In contrast to annual covercrops, the trees, if they can withstand the continued cutting, can be long lived and, if deep-rooted, drought tolerant, and even capable of conveying deep-lying nutrients to the upper soil strata.

The idea here is to provide a range of facilitative services at a low cost, using perennial woody or non-woody species, with less tree-on-crop competition. A proviso is that the understory not get in the way during planting or harvests. For this, the crops are almost always perennial.

One documented application, coconuts over the truncated tree gliricidia increased nut yields by 12%. In an exception to the perennial crop rule, seasonal maize planted over the pruned woody plant sesbania

PHOTO 5.3. A crop over tree system, in this case, nitrogen-fixing tree species beneath bananas and oil palm mix. This photo is from central Africa.

increased maize yields. This, coupled with salable firewood, made this an attractive revenue-oriented system.

Increased yields and firewood are not always the driving motivation. Having a slow-growing, weed-suppressing perennial in the understory can cut costs. Cost-oriented versions do exist, when applied to a rubber plantation, a woody shrub as an understory dropped weeding costs by 1/ 3.[22]

This agrotechnology is, in large part, a substitution effect, the understory adding or displacing an input. In the above cases, two added nutrients and one reduced labor. Economics will focus on primary effect, not forgetting tree pruning costs or any harvest constraints imposed.

Physical Support Systems

Following nature's plan, it is possible to substitute strong-stemmed perennials for vine-supporting trellises. The vines, in seeking sunlight, grow atop other plants. If done at a low cost and with accompanying salutary ecological dynamics, this can produce favorable economic outcome. Variations, vines above, in, and below tree canopy and the height of the tree, pruned low or tall and straight, are all part of the design.

With strong support-vine complementarity, yield increases should be expected. One study found that yams, grown upon the pruned woody perennial gliricidia, doubled in yield when compared against unsupported plants.[23] Despite complementarity yield gains, the replacement of expensive artificial trellises should be more of a cost gain and a form of cost orientation. As a large category, the variations of support systems are further discussed in Chapter 14.

Shade Systems (Heavy)

Exploiting the fact that some perennial tropical crops, e.g., coffee and cacao (cocoa), are shade resistant, the defining and operational characteristic of a heavy shade system is a thick canopy over the crop. The shade component can be one or many species. The resulting sun-screened environment offers a constant micro-climate, improved nutrient cycling, and well suppressed weeds.[24] Since the goal is reduced costs attended by reduced outputs, these systems are almost always cost oriented. There are three main versions; (1) a natural forest canopy, (2) purposeful shade species, or (3) a wood-producing plantation canopy.

The first of these, a canopy of natural forest trees, is the cheapest to implement, often requiring no more effort that planting beneath a natural forest. Besides the crop, there are added economic possibilities in harvesting natural forest outputs, including selective logging.

A purposefully planted shade species offers more market flexibility. The economics here involve balancing the amount of shade against losses in output and revenue. Light brings higher yields, but with greater costs (i.e., more weeds) and reduces sustainability (i.e., less capture and an increase in the outflow, through harvests, of mineral nutrients). Too much shade harms yields. Somewhere, as determined by the shade level, there is an optimal profit. Flexibility comes as the shade trees can be branch thinned when commodity prices are high, allowed to grow back when prices sag. For the facilitative tree, fast growth is one of many DPCs, others include fixing nitrogen, not harboring harmful insects, etc.[25]

The third version, having a wood-producing plantation overhead, is the least utilized. A high value hardwood, destined for harvest at some far-future date, would provide a substantive monetary reward at the end of the plantation cycle. Wood sales would cover the destruction and replanting of the primary crop. The issues, besides the inability to respond to commodity price rises, are in the startup phase where accommodations must be made to shade the crop until the slower-growing, higher-valued trees can assume this task.

Whatever the form, heavy shade systems are almost always cost oriented. These verge on being, or with a natural canopy are, ecosystem governed. Optimal solutions are often within easy reach being mainly centered around a suitable companion tree and the branch pruning, increased light option.

Outside of measurable yields, costs, and risk, a host of tangible, incalculable gains occur in whatever version is utilized. One is the cooler, shady working experience, nice in a hot, humid climate. Also, relatively small areas provide a refuge for native bird species and small animals. Large areas of shade can do much to promote favorable hydrology and to flatten temperate extremes in nearby communities. This is also good for water conservation, furnishing streams and rivers with pure water with the proviso that fertilizer applications, if any, are well regulated.

ENDNOTES

1. The buffering of soils is a special case not discussed in depth. Examples on how soil characteristics, through biodiversity, are made more suitable for agriculture are in Dagar *et al.* (1995).
2. It should be mentioned, at least as an aside, that the main substitute may be the acquired knowledge and skills of the farmer (through adroit agroecology) when replacing agrochemical inputs (Gollin *et al.*, 2005).
3. The concept of DPCs has been touched upon by many, e.g., Beer (1987), with formal development in Wojtkowski (2002, p. 107).
4. There are lots of small-leafed cover species and descriptions thereof. Of those mentioned, crown vetch with sorghum and cotton is from Sainju *et al.* (2005) while rye grass or white clover with pepper or okra is from Biazzo and Masivnas (2000).
5. Insect-repelling species are listed in Chapter 4, the corresponding lists for non-productive species are much longer. For weed-suppressing mulch, examples are in Kamara *et al.* (2000), Chou (1992), and Jobidon (1992).
6. The cowpea/ tomato intercrop is from Roberts *et al.* (2005).
7. The $CER_{(LER)}$ is utilized only when the RVT equals the LER.
8. In comparison with the hypothetical example in Chapter 2 (Endnote 11), both are cost oriented (with an EOR of −0.22 vs. a stronger −6.24 as in this example). This difference would play out in application, the latter being utilized in very rural areas where land is more abundant, labor and other inputs scarcer.
9. Lowering the cost floor is a strategy for many less-valued commodities. Another example, perennial sugarcane and grasses raised for ethanol production would be more profitable than equivalently planted seasonal crops.
10. Whereas PPCs are scarce in the literature, cost versions are non-existent. As an abstraction, these do have value, especially as an analytical tool to advance cost-oriented agroecology.
11. A discussion of biomass and the interval of application is in Makumba *et al.* (2006).
12. The complexities of weeds and intercrops are brought to the fore in a study by Poggio (2005).
13. Although not all covercrops produce economically attractive levels of insect control, most can if the right prescriptions are followed (see Altieri and Nicholls, 2004, p. 123).
14. The dehesa system can also incorporate cork oak trees, thereby reclassifying this under agrobiodiversity.

15. Live fences, those serving to protection from or to contain wild and domestic grazers, are covered in Chapter 9.

16. Roots and root spread are discussed in Schroth (1995).

17. The high LER value, 3.69, was presented by Ong (1994). Reynolds (1991) found 40% yield increases more representative. More information on the economically mixed economic picture for alley cropping is in Nair *et al.* (1999) and Raderama *et al.* (2004).

18. This study, which found five bad, one good tree-tea pairing, is from De Costa and Surenthran (2005).

19. The permanence of hedgerow alley cropping has been addressed. Some have utilized pigeon pea as a short-term substitute, but this carries with it a need for replanting every few years.

20. Firewood harvests tipping the economics toward hedgerow alley cropping is a conclusion by Swinkels and Franzel (1997). As an added note, because of poor-nutrient dynamics, undecayed wood is not useful as a soil additive or a soil cover. Very small branches, having a high percentage of bark, can be utilized as a ground cover.

21. The mini-hedge design is presented in Cooper *et al.* (1996).

22. The coconut–gliricidia application is from Liyanage (1993), the maize–sesbania intercrop from Sanchez (1995), weed control rubber by Budelman (1988).

23. Examples of live supports are in Salam (1991), Cálix de Dios and Castillo Martinez (2000), and Chesney *et al.* (2000). The yam case is from Budelman (1990).

24. Recent authors discussing aspects of heavy shade are Perfecto and Armbrecht (2003), Perfecto *et al.* (1996), Mas and Dietsch (2004), and Rice and Greenburg (2000).

25. Beer (1987) suggested some DPCs for heavy shade trees, others, e.g., Akyeampong *et al.* (1995) and Faizool and Ramjohn (1995), have added to this list.

6 Temporal Economics

Well accepted, the rotational vector is a strong tool in agroecology. Far from new, ancient farmers saw advantages in seasonally changing the content of agricultural plots, doing so with or without a fallow period. This was a powerful component in early-period agriculture and remain so to this day.

Rotations are not always seasonal. The rotational periods for orchards and many agroforestry systems can span decades. Within these long rotations, there can be sub-rotations. For these, the ecosystem remains intact as the individual species rotate in and out. The longest reigning systems are permanent but still affording sub-rotations for the individual plants.

APPLICATIONS

In nature, there are established successional sequences. These generally commence after a natural disaster, e.g., a severe fire. The farm version is not an exact duplicate of what nature does. Instead the sequence often starts anew after a bare ground phase. Although most farm plots never leave the first successional stage, one crop can still set the stage, through soil conditions, for the next. Along the way, setting the stage for other positive ecological tasks.

As in most of agroecology, economic success comes more by following, not fighting, the patterns set by nature. In general, longer rotations tend to be more cost oriented. Continual agrosystems, those that do not require replanting, have the potential to be, and often are, the most cost oriented.

UNDERLYING BIODYNAMICS

Rotation, as one of the more potent vectors, can and should do a lot. The existing guidelines are fairly strong and cover what should be the majority of cropping situations. Rotations, full ecosystem or by way of individual species, are expected to:

- rejuvenate the soil through increased mineral resources;
- interrupt the life cycles of damaging insects and diseases and resource-demanding weeds;
- curtail the accumulation of plant-toxic, allelopathic compounds, including decay-retarding wood extractives, left behind by some plants; and/ or
- aid in retaining water and countering drought by laying down moisture-holding, in-soil biomass.

Yield Gains

Rotations are not always necessary. Within-season, well-balanced, interplant dynamics of a well-planned agroecosystem may capture all the essential resources needed for a given season. In this case, cross-seasonal thinking or planning is not required. This is more common with cost-oriented, long duration or permanent tree-based agrosystems where the harvests and the removal of nutrients balance favorably with the nutrients internally captured.

Closed season dynamics are not always possible, especially where crops are seasonal and harvests are intense, removing a fair percentage of the one or more nutrients. When this happens, one seasonal crop should set the soil conditions for the next, i.e., the soil nutrient content after one crop is harvested is right for the next crop species. Ideally this should continue across many crop species on into the distant future. If this cannot be so designed, fallows, inserted into sequence, should amend any nutrient shortfalls.

Temporal DPCs

The most telling dynamics occur with full rotations and a harvest-ending cropping season. After the removal of a crop, intercrop, or some crop-facilitative species mix certain soil conditions linger and, unless the species (one or more) is a usurper of all mineral resources, some useful elements remain. All the better if one crop has the ability to attract and retain one or more nutrients. If the case, the soil will be in good stead for the upcoming crop. This is in addition to the constant influx of air-borne nutrients experienced by most sites.

Just as plants have desirable plant characteristics (DPCs), the residuals they leave behind constitute a temporal DPC. Two crops are interspecies compatible because they have complementarity in their DPC, one may seek more nitrogen, less phosphorus, the other may want more phosphorus, less nitrogen. On a site rich in these minerals, both can thrive and yield.

These same relationships can hold across seasons. A phosphorus-demanding species may be followed by a species that uses this element in small amount. Cross-seasonal complementarity is exploited in much the same way as intraseasonal complementarity.

The advantage is that water and light, two often limiting resources, do not constrain temporally separated species. Take the classic bean and maize intercrop. It is recognized that, if seasonal rainfall is below 600 mm, this combination will not flourish. Where rains are below this lower limit, the strategy is to plant these in alternate seasons, still taking advantage of known complementarity.

This has broader applications. Rather that focusing on the before and after soil properties (the temporal DPCs), harnessing known interspecies dynamics offers a shortcut in finding rotational sequences.

In many regions, time-tested rotational sequences are well established. An example from medieval Europe alternates three crops plus a fallow. This starts with winter wheat or rye, following this comes spring barley, oats, or buckwheat. Next is pea, broad bean, or vetch, all nitrogen fixing. The final step before repeating the sequence is a fallow that doubles as a pasture.[1]

To avoid overlooking potential sequencing plant partnerships, the before and after soils method (using temporal DPCs) is a viable alternative. This involves, through soil testing, matching the residual left by one crop with the needs of another, temporally stringing assorted seasonal species until soil conditions exist to start the sequence anew. If closing this circle proves difficult, fallows do restore soils and can close sundry nutrient sequencing gaps.

Fallows

The fallow is the temporal equivalent of facilitative biodiversity, i.e., fallows are non-productive periods that leave essential elements in abundance for the upcoming agroecosystem. Much can be expected, especially if a fallow has no productive role and can be fully committed to the ecological tasks as assigned. Foremost are the yield gains from post-fallow rejuvenated soils.

As expected, this is not a simple topic, options abound. As many are highly species and site specific, only a few options in a long list are discussed.[2]

Often, in the interest of cost control, naturally occurring species populate a fallow and can refurbish the soil for the next planting. More intense and of greater utility are planted fallows with high nutrient gathering ability. For maize, a planted fallow of a one, two, and three season length provided after-fallow land equivalent ratio (LER) values of 1.4, 3.1, and 3.3, respectively. In another case, a planted fallow of 18 months was found to provide the equivalent of 500–600kilograms per hectare of nitrogen.[3]

Another possibility is to plant with the express purpose of expanding upon the facilitative gains. For this, a mix of species can be used. Some might fix nitrogen, some (e.g., marigolds) might reduce nematode populations, others will enhance in-soil biomass. It might help if potential weed species are not included although weeding a planted fallow is not an option. A species-preselected fallow mix is a more-the-merrier approach, i.e., more species are generally better than a few. Directed biodiversity in fallow planning is a topic not yet full explored.[4]

There are pros and cons in the pre-planting burn. More nutrients are immediately available after a fire, but many are lost through combustion. Additionally, insects, good and bad, are adversely affected and some weed seeds are destroyed. For no-burn systems, the nutrients are gradually released and more are available. The disadvantage is that these may be bound in undecayed biomass during the peak crop-demand period. As a decision variable with few formulated rules to go by, decisions along these lines are site and crop based.[5]

A key provision for a planted fallow is the cost. It is a fairly simple cost–benefit analysis, the costs of an intensely managed fallows must be kept well below the benefits obtained, the benefits being yield increases or the fallow as a substitute for ex-farm chemicals, e.g., fertilizers, insecticides, etc.[6]

Cost Reductions

Outside of fertilizer savings of rotations with and without rotations, there can be some very strong cost gains in the form of weed control. This can be a good strategy when a specific weed species is targeted. The weed, striga, is parasitic on maize and sorghum, providing a false host can prematurely trigger this weed, leaving fewer seeds and weed growth when maize or sorghum reoccurs in the rotation.[7]

Another hard to manage weed is African spear grass. Difficult to control when amongst monoculture of a staple crop, it is easier to handle with the aid of other species. The strategy is to first plant a control species, usually during a fallow, later the crop.[8]

Less explored, but still possible, are the temporal DPC of particular weed-combating species. Wild mustard or *Brassica* spp. (winter rape,

B. napus, or spring colza, *B. campestris*) are among those species that are weed toxic, the hope being that the residual effect can linger with a future, positive gain.[9]

As a full rotation control, established sequences do eliminate weeds as a economic factor. Wheat, corn (maize), prozo millet, in sequence, followed by a fallow does accomplish this. It is not always this simple. Control of the most detrimental weeds does require a schedule that includes the type of tillage, management of the crop residuals, as well as the rotational setting.[10]

Risk Abatement

Rotations do reduce risk. The condition of the soil, including water-retaining capacity, is part of this. A number of mechanisms are in play.

Climate

Although climatic moderation is not often associated with rotations, this is one indirect, but important exception. Low rainfall and high transpiration can be countered through moisture-holding, in-soil biomass. Crops that leave a large amount of residual biomass which, once plowed under, can absorb and hold moisture. This in-soil biomass has the potential to sustain an upcoming crop when rains are heavy, but infrequent.[11]

Insects and Diseases

Some species kill off unwanted insects. Clearly, in leaving a site free from this scourge, gains occur. It is not uncommon for insects and diseases to wait for the return of their preferred crop. A lengthy delay, as when other crops or a fallow period follows, can out wait and eventually dispatch an awaiting plague. There are many examples, growing seasons without barley or a related species can reduce instances of diseases in subsequent barley crops.

The delay is not the only mechanism. Plants use natural toxins to ward off harmful organisms and, at times, these do linger in the soil. For example, rather than simultaneous biodiversity to counter yield robbing nematodes, it might be better to plant nematode-countering plants in the season before the susceptible crop. Varieties of marigolds are a well-known nematode counter with a cross-seasonal residual. A pre-seasonal strategy might hold if the marigolds are complementary with the crop planted the year before the susceptible type, but not yield compatible with the crop needing protection.

The planting of facilitative species the season before it is required, relying on a lingering in-soil effect, constitutes a temporal specific interaction zone. This expands the temporal options and, as exampled, can prove particularly useful if a facilitative species is over competitive with one, but not all the crops in a sequence.

Part of a temporal sequence might include fire. As with fallows, there are nutrient questions and whether non-burned biomass has continued worth. In burning crop residues, some insect pests are destroyed, e.g., the alfalfa weevil can be controled by burning the post-harvest residual alfalfa. To the contrary, fire is an indiscriminate killer that can negatively effect desirable predator insects, thereby disrupting a general control strategy. Before being considered, the situation should be well understood and the targeted insect known.

Relevant Guidelines

Rotations offer a range of guidelines. The general rule that if two crops interplant well, i.e., the intercrop LER values generally exceed one, these will also do good, or better, in rotation. There are caveats, including the need for set or predetermined sequence.

Fallows do increase cropping flexibility, especially if a sequence of low-valued crops is needed before restarting with the high worth species. There are other options. The temporal DPC is a direct offshoot of DPCs and is applied in much the same way, as a component in an insect, weed, or other control strategy.

ECONOMIC MEASURES

As expected, the LER can be the basis for rotational analysis. There are two questions in the temporal sphere: (1) sustainability and (2) the best sequence.

Sustainability

Since rotations are the most manifest expression of the temporal plane, sustainability may be best offered under this heading. The concept is not well defined,[12] the simplest definition, that of the long-term productivity of the land, may prevail with a statistical trend determination as the best indicator with a temporal LER addressing the apples–oranges issue. The disadvantage of trend analysis is the need for a long data stream.

The calculation, that of the LER for the time periods, is done through statistical regression analysis. The better scenarios are a flat or upward trending regression line, the worst tend downward. This statistical method has one all-inclusive advantage, unplanned events, e.g., insect attack or drought, can be factored out, that is if enough data is available. If not, knowing that land will continue to yield, at predetermined levels, should be enough to continue.

Ordering

The cropping ordering presents two options: (1) a comparative RLER and/ or (2) absolute temporal LER (tLER).[13] As comparative RLER is formulated as

$$\text{RLER}_1 = ((\text{LER}_a)_1 + (\text{LER}_b)_2 + (\text{LER}_c)_3 + (\text{LER}_d)_4)/4 \quad (6.1)$$

Species a through d, in this order, are in the rotational sequence. The seasons are numbered 1 through 4. Each seasonal LER is calculated using, as the LER denominator, the yield when the respective crop is not part of any cropping sequence, e.g., $(\text{LER}_b)_2 = (Y_b)_2/(Y_b)_1$. From this, the above equation can rewritten as

$$\text{RLER}_1 = (1 + (\text{LER}_b)_2 + (\text{LER}_c)_3 + (\text{LER}_d)_4)/4 \quad (6.2)$$

A simple comparison of the order will show which lineup is best.

The other method, less comparative, more unconditional, is to utilize a baseline figure. The baseline is found through the LER of one crop without outside inputs as

$$\text{Baseline LER} = (1 + ((Y_a)_2/(Y_a)_1) + ((Y_a)_3/(Y_a)_1) + ((Y_a)_4/(Y_a)_1))/4 \quad (6.3)$$

Y_a is the yield for one crop (species a). This is sequenced across four seasons where comparison are made against the first season (season 1). The tLER is calculated as

$$\text{tLER} = \text{RLER}_1/\text{baseline LER} \quad (6.4)$$

The following numbers demonstrate a sequential evaluation:

	Season			
Yields	*1*	*2*	*3*	*4*
Crop a	1200	1150	920	850
Crop b	475	390	350	310
Crop c	1525	1370	1225	1160
Crop d	970	920	860	822
a–d in sequence	1200	450	1325	930

$$\text{RLER}_1 = (1 + 0.94 + 0.86 + 0.95)/4 = 0.94$$

$$\text{Baseline LER} = (1 + 0.95 + 0.77 + 0.71)/4 = 0.86$$

$$\text{tLER} = 0.94/0.86 = 1.09$$

This example indicates that the a through d cropping sequence is site superior (1.09 > 1.00) when compared against continued monocropping of the primary species (crop a).

Also possible are other comparative measures. Adding selling prices to the above will result in a temporal relative value total (tRVT):

$$\text{tRVT} = ((P_a(Y_a)_1) + (P_b(Y_b)_2) + (P_c(Y_c)_3) + (P_d(Y_d)_4))/(P_a((Y_a)_1 + (Y_a)_2 + (Y_a)_3 + (Y_a)_4)) \quad (6.5)$$

For this, species a is assumed (through price times yields) the most valuable. Continuing with the numeric example, assumed selling prices are, for crop a, \$0.75; for crop b, \$0.82; for crop c, \$0.35; and for crop d, \$0.67:

$$\text{tRVT} = ((\$0.75)(1200) + (\$0.82)(450) + (\$0.35)(1325) + (\$0.67)(930))/(\$0.75(1200 + 1150 + 920 + 850)) = 0.75$$

There are cost savings associated with rotations. These are reflected in a tCER. The equation might be

$$\text{tCER} = (1 + ((C_a)_1/(C_b)_2) + ((C_a)_1/(C_c)_3) + ((C_a)_1/(C_d)_4))/4 \quad (6.6)$$

The assumption in having $(C_a)_1$ as the numerator is that the seasonal inputs remain constant with no interseasonal savings, otherwise the yearly changes in cost would be utilized (e.g., $(C_a)_2$, $(C_a)_3$, and $(C_a)_4$). Using the ongoing example with assumed costs:

$$\text{tCER} = (1 + (500/300) + (500/250) + (500/375)) = 1.27$$

More revealing is the $\text{tCER}_{(\text{tRVT})}$. This is computed as

$$\text{tCER}_{(\text{tRVT})} = (\text{tCER})(\text{tRVT}) \quad (6.7)$$

Numerically exampled, this is

$$\text{tCER}_{(\text{tRVT})} = (1.50)(0.75) = 1.12$$

Also of use is a temporal economic orientation ratio (EOR).[14] This is easily derived as

$$\text{tEOR} = \text{tRVT} - \text{tCER} \quad (6.8)$$

and exampled as

$$\text{tEOR} = 0.75 - 1.12 = -0.37$$

which indicates a fictional rotational sequence that is mildly cost oriented and should be applied accordingly. After checking the numbers through financial accounting, such a sequence should produce positive economic results.

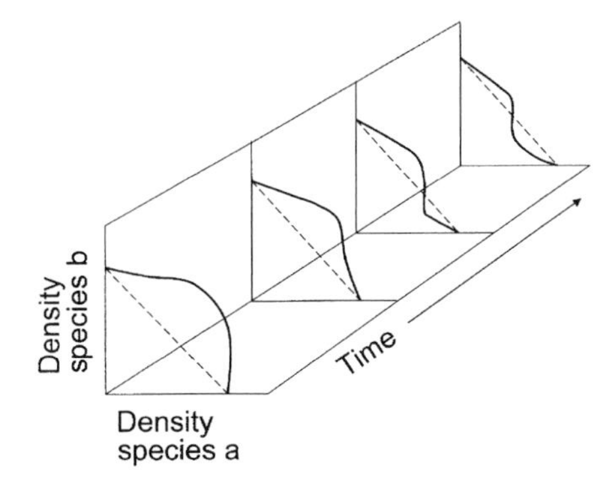

FIGURE 6.1. The effect of time and growth on the PPC. In this drawing, the PPCs are shown for representative points along a timeline.

Temporal PPC

As crops grow, the intercrop relationships, via the essential resources, changes. Within a season, a taller species can over top a shorter one, robbing sunlight and changing the resource balance. The effect and outcome is more pronounced interseasonally where, as trees grow and their canopies enlarge, they may eventually subdue an understory.

This can be expressed as an temporal production possibilities curve (PPC). Figure 6.1 shows, in a hypothetical case, how the PPC transforms over time. In this illustration, species b (vertical axis) gradually suppresses the yield of species a. This concept generally applies to long duration multi-species systems.

This analysis does work for optimizing rotations by optimizing the spatial pattern, species ratios, and the spacing of seasonal intercrops with regard to soil conditions left from the previous planting. Use depends on data availability.

Net Present Value

The common temporal financial expression is the net present value (NPV). This one-number, monetarily expressed convention does offer intuitiveness and comparability. However, NPV does present a short-term bias. For the latter reason, it is not advocated as a sustainability measure. Generally, the applications are purely financial, e.g., deciding if long-term tree rotation, wood or fruit, is a better investment compared to an interest-bearing bank account.

Despite use, many applications fall outside the scope of NPV analysis. The latter can occur when farmers plant trees to cover future expenses, e.g., wedding dowries and college funds for newly born children. This is

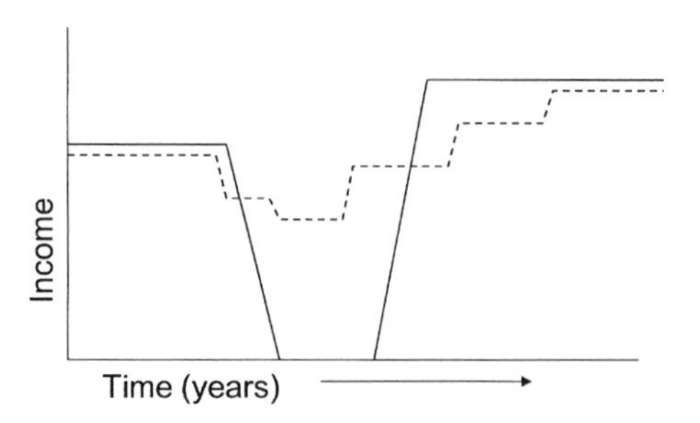

FIGURE 6.2. A representation of income gap (solid line) and what might happen if measures are taken to reduce the severity of the gap (dotted line).

the case when banks or other financial institutions are not an option or if there is no ready cash on hand to deposit.

Another analytical dangers are encountered when commencing long-term rotations where initial returns are some years in the future. This occurs when, as mentioned, converting from seasonal agriculture to farm forestry. NPV may initially show the transition as unwise, in part because of the income gap. Once a long-term sequence of tree harvests is underway, this may be far more profitable than the seasonal alternative (as in Figure 6.2). Income gaps are discussed in greater detail in Chapter 15.

Except in the most stringent of commercial settings, a single NPV figure may not be the best indicator. It is often wise to take a deeper look at the situation, including any increases or reductions in risk from a long duration crop.

DESIGN VARIABLES

Some design variables magnify the cross-seasonal possibilities. Included in any list are frequency of planting and the timing of plowing and sowing.

Delayed Sowing

A subtle, but valid temporal expression lie within one cropping season. Weeds, being opportunistic plants, often emerge at the first sign of a favorable growing conditions. A late plowing and the sowing of a shorter-season variety may circumvent this threat.

Some have suggested multiple plowings, the second occurring after the weeds have sprouted. This may be viable option if cost of plowing is less than the cost of weeding or the cost of herbicide application.

A late planting can do more that manage weeds. Yellow dwarf virus and some weeds are controlled by planting barley very late in a season.[15]

Planting

For the planting of crops and trees, there are a number of methods. Listed, these are (a) seeds, (b) seedlings, (c) striplings, (d) stumps, (e) stem, and (f) series. Only the first two are found with seasonal crops, the others (c, d, e, and f) are for use with woody plants. Success depends on the woody species, e.g., pines can only be established as seeds, seedlings, or striplings. Other tree and shrub species offer more transplant flexibility.

Seed planting is the agricultural standby. The seeds can be row planted with concerns on the inter- and intrarow species. The broadcasting of seeds removes the row measurements and, under the assumption that they are uniform spread, makes density the main criteria.

There are temporal aspects that aid intercropping. It may be advantageous to stagger the planting of an intercrop, cheaper if all are co-planted. For example, the planting of a deep-rooted species, e.g., beets or carrots, should proceed, by a few weeks, the planting of a shallow-rooted plants, e.g., wheat or barley.[16] A seed coating, one that slows germination, may prove the better option, allowing for a less costly, simultaneous planting.

Seedlings or sprouts are newly germinated plants that are raised elsewhere and manually transplanted. The advantages are, due to the advanced age of those being repositioned, a shorted cropping period and a pre-transplant plant selection as only the more vibrant of sprouts are transplanted. This strategy is mandatory when the actual growing season is shorter than the productive life of a plant or more seasons in a single year increase overall plot profitability. The latter is often found with rice. Flooded paddies are a valuable resource and use is maximized if transplanting can give two or three, instead of one yearly harvest.

Striplings are very tall seedlings, instead of being around 20 cm, these are 2 m in height. When replanted, the lower branches are trimmed, leaving only a small canopy at the very top. The purpose, and main use, is to keep the canopy out of grazing harm.

Stump planting is a tree or shrub establishment technique. For this, the woody perennial is nursery raised to a diameter of about 2–4 cm. The plants are pulled from the ground, the roots cut-off to a length of 10–15 cm, the aboveground stem should be trimmed to a 3–5 cm height.

The gains in stump planting lies in reduced transport, handling, and transplanting costs. One person can carry many of these small stumps without fear of damage. For planting, a hole is poked in the ground, the stump inserted. If well done, establishment rates can be quite high.

Stem planting involves cutting a portion of a branch and placing this in the ground. This is a cheap method, the planting stock are the easily available branches of existing trees. The disadvantage is the high failure rate.[17] It may require a second planting, a season later at an added cost, to fill in the gaps.

A second form of stem planting takes this to a large dimension. Trees are nursery raised to a 10 m height, the roots and branches are removed, the long stem is planted. The gains here are with an established, superior stem form, i.e., tall and straight, and a greatly reduced time to maturity. This is especially important when occupying valuable land.

Series or progressive planting is highly limited, employed to fill gaps in hedges or fence lines of woody perennials. If a few die, a branch from a neighboring plant is weighed to the ground, e.g., with a large rock, and allowed to root in place.

ASSOCIATED AGROTECHNOLOGIES

The temporal plane does offer some distinct possibilities. Some involve variations of rotation dynamics, others take advantage of niches or available space with cropping sequences. There is another aspect, the phases of each would be a single non-temporal agrotechnology or the economic objectives might be better served if the temporal phases are different non-temporal agrotechnologies. For example, seasonal monocropping might be followed by seasonal intercropping.

Within this framework, the associated temporal agrotechnologies are:

- Single rotations
- Series rotations
- Overlapping cycles
- Taungyas
  - Simple
  - Extended
  - Multi-stage
  - End stage
- Continual

Single or Discrete Rotations

With single or discrete rotations, each seasonal land use is considered nutritionally unto itself, i.e., with no carry over or planning with regard to the future land use. Any deficiencies, whether these be soil nutrient shortfalls, carry-over insect, or weed problems are immediately handled through (a) appropriate ex-farm inputs (e.g., fertilizers, introduced predator insects, insecticides, etc.) or through (b) agrobiodiversity or biodiversity where interspecies dynamics insure enough nutrient capture to provide economically viable harvests (Figure 6.3).

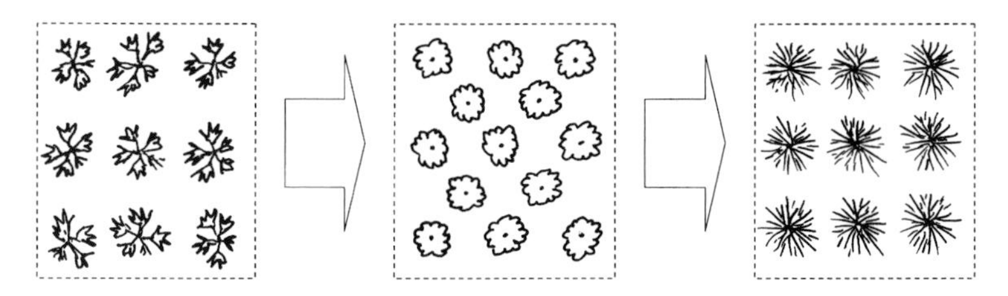

FIGURE 6.3. Discrete rotations with no cross-seasonal planning.

Single rotations offer a high degree of economic flexibility in responding quickly to changing market demands. Seasonal flexibility is traded against the ecological gains from well-planned crop successions. This may not be an obstacle if the ecosystem is biodiverse and self-sustaining, problematic with nutrient-draining monocropping. Because of the nutrient drain and the requirement for ex-farm replacement, these sequences are mostly revenue oriented.

Series Rotations

With planning comes the ability to utilize natural temporal dynamics. The theoretical underpinnings presented in this chapter underwrite the biodynamics in substituting residual in-soil effects for ex-farm or other inputs.

The fallow, resting the land for a brief or not so brief period, can be part of this. Harmful insects, those anticipating an edible crop and some plant diseases cannot lie in wait forever. Meanwhile soils continue accumulating nutrients. In one representative case, rotations, inclusive with a facilitative species, increased maize yields by 50%.[18]

The economics are clear. Cost goes down if lesser amounts of agrochemicals, i.e., fertilizers, insecticides, and maybe herbicides, are applied. Also, the per-unit costs of harvest can be decreased if yields remain high. The result, these sequences are more often cost oriented.

Overlapping Cycles

Given the right climate, overlaps in cropping sequences can be exploited. In an already mentioned case, tomatoes are interplanted with lettuce, after the lettuce is harvested, the tomatoes remain. This is not the only documented case. Maize, millet, and sweet potato can be intercrop, after harvesting the maize and millet, the sweet potato remain as a monocrop.[19]

This can be carried to a higher biodiversity plane by adding more species. Applications are more pronounced in the tropics where growing seasons are prolonged or never ending (Figure 6.4).

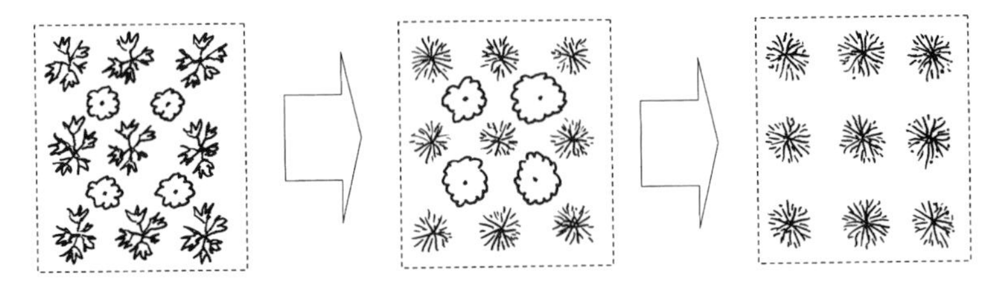

FIGURE 6.4. Overlapping cycles where one or more species from one season continue on into the next season.

Being the temporal dimension to multi-output intercropping, these address the same socioeconomic scenario. Overlapping cycles are best when farmers have little land, a comparatively large amount of labor, and many marketing possibilities. This occurs frequently near population centers where revenue orientation, high outputs for high inputs, become the norm.

Taungyas

When trees are part of a long-term planting, taungyas find use. There are four categories under this agrotechnology heading: (1) simple, (2) extended, (3) multi-stage, and (4) end stage. With a taungya, the tree is the primary species, the seasonal crop the secondary species providing output and ecological and economic benefits. Often the latter stage, if not a monocrop, ends with grazing, grasses and herbaceous forage being the secondary species.[20]

The taungya possibilities are limited, or may not be possible, when highly crop-competitive trees, such as pines or eucalyptus, are the primary species. Other than this, favorable economics are expected as weeding, are replaced by income-generating crops and the costs of other activities, e.g., soil preparation and the addition of nutrients, are prorated to more than one output. In general, taungyas, of whatever category, represent a shift toward revenue orientation.

Simple Taungyas

Taungyas were first formalized in forestry. The idea being that, between newly planted trees, there is much open space where essential resources go unused. Rather that having these claimed by weeds, at a removal cost, the space and resources could be occupied by a revenue-generating crop.[21]

This concept is valid for newly established orchards and treecrop plantations as the recently planted trees are small and draw few resources. Because weeds are missing and the land experiences an

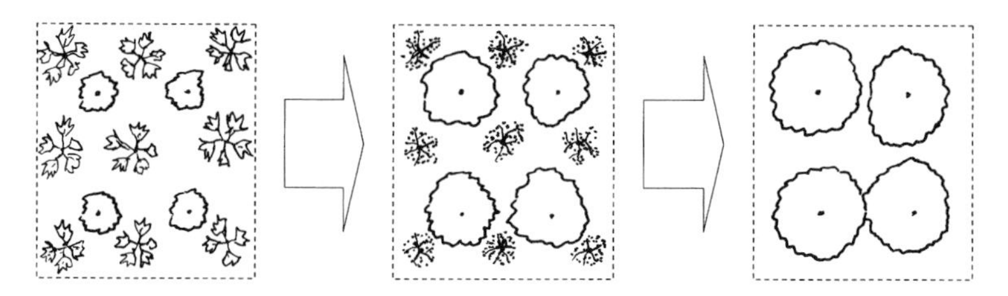

FIGURE 6.5. A simple taungya where a tree species, either fruit or wood producing, is grown with understory crops until the overstory canopies touch.

infusion of crop-directed nutrients, it is not unusual for the primary crop (the tree) to experience accelerated growth.[22] The simple taungya usually ends when the orchard or plantation becomes economically active, i.e., when the trees start to bear fruit or other outputs in volume (Figure 6.5).

Because these can be implemented with little effort, the simple taungya is a yes-or-no economic question. With a suitable crop, this becomes an easy yes. If needed, an outside farmer can undertake the cropping component (see multi-participant systems, Chapter 15).

Extended Taungyas

Expanding upon the simple taungya, it is possible to have crops under the trees (including orchards and treecrop plantations) throughout the tree rotation. There is a requirement for a cropping sequence brought about because the essential resources *vis-a-vis* the trees changes seasonally as the trees become increasingly competitive.

A number of examples have been noted. In South America, orange trees planted with maize, afterwards orange is paired with papaya. The papaya is replanted for as long as this crop remains profitable. From Chile, poplar trees are co-planted with maize or sugar beets, then oats or currents. A shaded pasture below the trees completes the 22-year tree cycle. Because the latter stage is part of revenue-oriented sequence, the orientation does not change despite this being, in essence, a heavy shade system.

The temporal PPC is useful with extended taungyas. There is no need for a yearly PPC, instead representative seasons would be used. In doing so, the extended taungya does past through a number of non-temporal agrotechnologies. The system starts agroforestry intercropping, may enter an alley cropping phase, going on to a light shade stage, ending with heavy shade. Each of these phases can be optimized. In the quest for optimization across phases and years, this might include the option to remove selected overstory trees.[23]

Extended taungyas are a difficult undertaking, requiring considerable planning and experimentation. There is a requirement for large areas. If too small, the crops grown during any one phase will not provide enough of a harvest to justify the investment in expertise and equipment. As a consequence, these systems are limited to large-scale commercial enterprises.

Multi-stage Taungyas

With orchards or treecrop plantations, the tree that starts a sequence may not end it. A mid-rotational shift need not be abrupt, instead it can be gradual as when the second stage tree is planted and matures before the first stage has ended.

The simplest type involves an overlap of two orchards, planting the second before the first is removed. Documented transitions include, from, Brazil, interplanted banana and cocoa. After the bananas are removed, fast-growing shade trees are planted. The cocoa, as an exception to the tree being the primary species, is the understory for the duration of the system.

The multi-stage taungyas are of limited use, most of which may be wood-producing or treecrop plantations.[24] Where applicable, these may represent a profitable shift away from simple rotation.

End-Stage Taungyas

In some forest-tree plantations, there might be a latter-stage thinning, leaving only a few scattered trees. These are expected to grow to a large size in a few years, with a corresponding increase in their market worth. With this thinning comes a surplus of internal ecosystem resources. These can be exploited for an end-cropping sequence or a pasture.

The end-stage taungya is a forestry opportunity often overlooked. These could prove valuable enough so that landusers change the harvest cycle of forest-tree plantations, both with farm and commercial forestry, to take advantage of this income source.

Continual Cropping

For many biocomplex agroecosystems, the temporal sequence is continual. For this, the individual plants change, but the overall agroecosystem lingers.

Sub-rotations, where individual species or plants come and go, is the common format when seasonal crops are involved. Continual cropping with seasonal species is only possible in tropical regions.

FIGURE 6.6. Continued cropping where, although the component species may change, the ecosystem, in form and type, remains across seasons.

Another example is agroforestry intercropping expressed as a mixed species orchard or a mixed species treecrop plantation. As not all trees are removed at the same time, the staggered comings and goings produce the sub-rotations in a continual agrosystem. Illustrated in Figure 6.6, these agroforestry versions are found in all climatic regions.

The most common continual systems, either tropical or temperate, are mix-species pastures and agrobiodiverse agroforests. Although much different in visual form, these have many ecological similarities. For starters, both have a large percentage of perennial species and a far lesser number of seasonal plants.

It goes without saying that pastures produce mostly animal forage or biomass for sundry uses.[25] In contrast, the output from agroforests is quite mixed. Trees or shrubs often dominate, but all sorts of products are possible. Fruits, nuts, root crops, spices, medicinal herbs, decorative plants, and wood poles example what is found on outputs lists.

Although revenue-oriented versions of agroforests do exist, most as near-city market gardens, the typical economic format is ecosystem governed and cost oriented. Sub-rotations are possible, e.g., to change the harvest mix in response to market demand or to control an introduced plant disease. However, this is labor intensive and seldom undertaken. As agroforests rely upon some unique biodynamics, discussion continues under complex agroecosystems (Chapter 14, section 'Complex Agroecosystems').

ENDNOTES

1. There is some disagreement on medieval agricultural practice, specifically if the main nutrient mechanism was cross-seasonal crop complementarity (as presented in the text) or the fallow. The former view seems prevalent in Vasey (1992) and Gras (1940), the latter in Slicher van Bath (1963, p. 64).
2. An interesting study on the many fallow options is in Kass and Somarriba (1999).
3. The maize-after-fallow LER values are from Kwesiga and Coe (1994), those on the amount of nitrogen from Ståhl *et al.* (2005).

4. In the beginning of this text, it was noted that, by laying out agroecology in a systematic way, gaps in practice are revealed. The idea of a mixed species fallow, one based on the DPCs of the individual species included, is one such unexplored avenue.

5. Some of the debate, slash-and-mulch or slash-and-burn is framed by Thurston (1997) and Peters and Neuenschwander (1988), respectively.

6. An example where the costs of a planted fallow were not in line with the benefits is in Langyintuo and Dogbe (2005).

7. Listings of striga control alternatives are in Ellis-Jones *et al.* (2004), Rao and Gacheru (1998) and Khan *et al.* (2000).

8. van Noordwijk *et al.* (1992) suggest this control for spear grass, eliminating this weed during a fallow, planting the crop before it has time to reestablish.

9. The influence of wild mustard is from Gliessman (1998, p. 121) and *Brassica* spp. from Grodzinsky (1992).

10. An in depth look at weed control, in which rotations are a key element, is in Anderson (2003, 2005).

11. The plight of having less in-soil biomass was faced and overcome by Versteeg *et al.* (1998).

12. Sustainability, as an agroecological concept, is discussed by Hansen (1996) and Stockle *et al.* (1994), problems of measurement by Beyerlee and Murgai (2001).

13. The temporal LER presented here differs from that found in Fukai (1993).

14. Although offered here, an EOR might find use at some future period when the economics have experienced greater development and it becomes possible to better assess comparative temporal values and connect these with optimal use situations.

15. The wheat, maize, millet, fallow as well as the elements in an anti-weed schedule are found in Anderson (2005, 2003). Others advocating integrated weed management are Liebman *et al.* (2001), Liebman and Gallandt (1997), and Altieri and Nicholls (2004, p. 69). Multiple plowing is an early suggestion by Worlidge (1669, p. 32). Virus control is also in Anderson (2003).

16. The timing sequence for the co-planting deep- and shallow-rooted species comes from Anon. (1872).

17. A failure rate of 40% for branch planting has been seen by this author. This contrasts with a 95% establishment for stump planting (also from first-person experiences). The exact rate for any planting would be species and rainfall dependent.

18. The 50% maize increase is from Eilittä *et al.* (2003).

19. The maize, millet, sweet potato intercrop is from Brown and Marten (1986).

20. With a long history of use, taungyas are mentioned, e.g., by Browne (1832), Schenck (1904), Blanford (1925), and Menzies (1988).

21. Crop sub-rotations within the simple taungya are possible. In a rubber tree taungya, Esekhade *et al.* (2003) found clear economic gains from a simple taungya whether or not rotations were employed.

22. One such study on accelerated growth is in Chifflot *et al.* (2006).

23. The optimization of a long-term plantation byway of bioeconomic modeling is in Wojtkowski *et al.* (1991).

24. Silvicultural applications for multi-stage and end-stage taungyas are found in Wojtkowski (2006b).

25. Other uses can be hay or hay bales for erosion control or biomass for ethanol production.

7 Genetic, Varietal and Locational

This chapter examines three somewhat related vectors: (1) genetic, (2) varietal, and (3) locational. Some, mainly the genetic vector, have received considerable attention. Still, each finds use and can contribute the agroecological whole.

The genetic vector brings, through domestication, crops from their wild, untamed state to a higher degree of usefulness. The gains have included larger and more numerous fruits, better per plant yields, ease of planting, ease of harvest, along with taste, storage and other post-harvest concerns. Since there is always room for improvement and the process continues.

The varietal and locational vectors involve finding, often amongst hundreds of choices, the correct crop and/ or variety for the right job, i.e., ideally matching each plant against local growing conditions. Although this need not be the case, use seems confined mainly to regions where there are many varieties of a single staple crop and farmers, through decades or centuries of exposure, understand how this resource is best applied.

These two vectors, varietal and locational, have much in common. Although they address the same problem, i.e., finding what grows and yields best on any given plot, these arrive at a solution in a different way. Because of this, these are not exclusive, e.g., a locational approach can be fine-tuned through a varietal process.

APPLICATIONS

The genetic approach has long history of use. The earliest agriculturists found useful plants in the wild and began the process of transforming them into something better for agriculture. The pace was slow and may have been done without conscious input. Progress only requires that each farmer replant the best from the previous years crop.

The green revolution, which began as a formal program in the 1950s, was predicated on making giant steps in improving the yields of common crops. The goal was a food surplus at the national level, not farm profitability or farm environmental objectives.[1] The most expedient approach was through ex-farm chemicals. The result, improved species that is best grown in monocultures, plants often less suitable for use in intercropping or other agrobiodiversity associations.

This is not the only genetic avenue. Others advocate for broader goals, breeding same varieties for large farms (as is done) but also varieties intended for the small-scale, specialized producers. The latter being undertaken with more of an eye toward plant–plant (intercropping) associations and/ or cost orientation.[2]

Taken to an extreme, genetics seeks to armor crops against all possible adversities. Some consider genetic modification as inordinate as this can involve taking genes from unrelated species and utilizing these in the quest for the 'perfect' agricultural plant. This has unleashed a degree of hostility for a number of reasons. In reaching outside the scope of nature, the perfect crop may be an entirely human construct. The worst case is an unnatural plant at odds with, or detrimental to, local flora and fauna much the same as when introduced species run amuck, reducing or destroying natural ecosystems.

Outside of this danger, not all can be planned for. Although well armored against adversities, nature has a tendency to find and exploit clinks in defensive armor. There is a risk of loosing large areas when an unconsidered adversity strikes or a genetic weakness is found by insects, fungi, or the like.

In commercial agriculture, storage, longevity, and transport handling often trump taste, nutrition, and food freshness. In breeding and marketing only a few varieties, there are concerns on the nutritional front. Different varieties of vegetables and grains contribute to human health in varying ways, e.g., one manifestation is fruit color. Eating better can mean consuming greater agrodiversity, not just in number of species, but in different varieties for each species.[3]

The varietal vector seeks, instead of developing a variety, to find one. For many crop species, there are a myriad of choices. For wheat, the figure was once put at over 600 available types.[4] Rather than seeking the perfect variety (a crop-first approach), nature presents other options.

The intent of the locational vector (the plot-first approach) is to match a crop, i.e., the essential resource and other need profiles, against the attributes of a site. If a plot is not suitable for maize, some other, more site suitable crop is substituted. If there is no choice but to plant maize, then a variety is selected that does best. In addition to immunity to diseases, inedibility to insect species, resistance to weeds, and other in-plot concerns. Attributes, such as taste, nutrition, and storage life, are also added to the mix.

UNDERLYING BIODYNAMICS

There is a body of theory, and ample practice, to support location as an appropriate vector. The theory holds that a crop type, wheat, maize, rice, potato, etc., have an inherent and preferred climate. This comes from the evolutional history of each crop. Many, if not all, of the agricultural (domesticated) versions will initially not fall far from the pre-agriculture prototype in their preferred ecological surroundings. User, requirements can force a change.

Examples abound, maize originated as a sub-tropical species, as varieties evolved, the ability to grow in ever-shorter growing seasons made this into a temperate crop. Latter developments, especially better cold resistance, have rounded out the process. The genetic process is tasked with:

- insuring high yields,
- resisting harmful insects and diseases,
- enduring drought,
- withstanding climatic extremes.

The result, certain tasks are expected for the varietal or locational vectors. These are to:

- maximize the use efficiency of all site-available essential resources,
- mitigate the negative effects of temperate,
- stand up well to negative soil conditions,
- help reduce insect and disease risk.

Varietal Selection

The ability of crops to accommodate site differences is graphically shown in Figure 7.1. The various varieties are represented as short vertical lines. Two crops, top and bottom, are represented. The horizontal axis can be one or many site attributes, i.e., in-soil nitrogen, soil moisture, air temperate, etc. If the horizontal axis were to represent existing or average moisture content, as symbolized by the dark vertical line, the lower crop would offer many more suitable varieties than would the upper crop.

In actual practice, the graph would not be one-dimensional (with a single attribute), but encompass any number of site attributes. Many

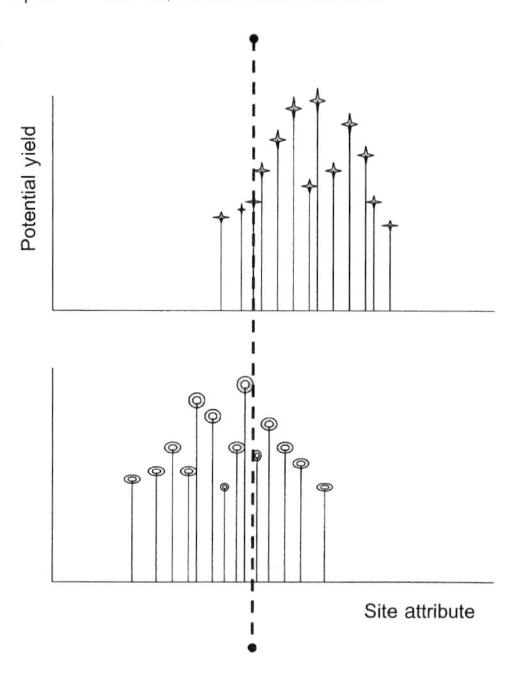

FIGURE 7.1. The varietal distribution and potential yields for varieties of two species (top and bottom) with regard to a site attribute (e.g., soil moisture or nutrient content). Each topped vertical line represents a single variety and its yield at its optimal point. For an existing site attribute (vertical dotted line), the lower species may be the better site choice.

regions, especially those where the crop originated, may have many varieties from which to choose. There may be tens, if not hundreds, of genetic variations.

This is why, in Figure 7.1, the height of the varietal lines, an indication of potential yield (vertical axis), do vary. With one variable, say soil moisture, each variety responds differently, some out yield others. The situation may be entirely reversed with another site variable, say phosphorus levels.

The core decision on which variety to use would be based on limiting resources, the threats, and a crop varieties ability to address each. After satisfying growth criteria, farmers then select based on other desirable properties.[5]

Locational Selection

If a farmer is to get the best result, it may be best if the crop chosen is well suited to a micro-site. This may require that the crop strips that contour a hillside vary in content (species or variety) in response to the elevational gradient.

This is the basis for locational selection. Figure 7.1, bottom, shows which crop species are best suited to an actual site attribute (the long vertical line). The closer the original pre-domesticated prototype to the site attribute in question (show as a long vertical line), the more varieties from which to select.

In comparison, the top crop (Figure 7.1), in having fewer varieties that match the attributes of the actual micro-site, may not be the ideal crop. If the lower axis were to represent moisture, the lower crop would be better adopted to an arid planting. Although a less documented approach, farmers do choose location as a cropping strategy.[6]

Relevant Guidelines

What is advocated is to first select the right crop and then choose the best variety. Some of this involves a high land equivalent ratio (LER), other aspects are in addressing and mitigating risk. Of course, these come with what can be overwhelming caveats. The desirable plant characteristics (DPCs) of the different varieties are seldom known, farmers seek to plant high value crops, seed suppliers do not offer large choices, and consumers may be reluctant to accept foods different from what they are accustomed.

ECONOMIC MEASURES

With the clear goal of match the crop or variety with the site or with the goal of specifically developing the fitting variety, determination of success, or at least an indication of being on the right path, is through LER. This can be as a monocultural expression, i.e.,

$$\mathrm{LER} = \frac{Y_{\mathrm{a2}}}{Y_{\mathrm{a1}}} \tag{7.1}$$

that has Y_{a1} as the yield of the variety in current use, Y_{a2} the proposed variety.

Some varietal insertions are destined as a component in an intercrop. For this, the LER would be in its original or near original form (as presented in Chapter 2).

DESIGN VARIABLES

Two of the vectors covered in this chapter, varietal and locational, represent two sides of the same coin. The difference, seemingly abstract, is real and is an agroecological tool.

Varietal

In the chapter on agrobiodiversity (Chapter 4), the varietal monoculture was discussed. This employs different, interplanted varieties selected because they confer a degree of protection against climate variations or against threatening insects or diseases. The theory of the plant prototype and the variation around a norm (often the wider the spread the better) explains how the multi-varietal monoculture is supposed to work; keeping in mind that other motives may underwrite multiple varieties (Photo 7.1).

Locational

Crops become popular, in part, because they are better suited to a climate. Rice and cassava are both tropical and grow well in these regions. A species does not have to be grown where it once evolved. Rice, of southeast Asian origin, as well as cassava, of South or Central American origin, is both commonplace in Africa.

There are pros and cons to exotics, there are cases where infirmities have not followed a plant to its new location. In Africa, rubber trees do not have the disease problems of those in native South America. In time, diseases and insects do infect well-traveled hosts. If the conventional plagues cannot journey in pursuit, nature has a tendency

PHOTO 7.1. Two side-by-side maize varieties. The use motive may be more for market or labor concerns, in having non-coinciding harvests, rather than to take advantage of site and soil differences.

to evolve new ones.[7] The constant movement across widely scattered plots does offer a passing reprieve. As such, travel is a weapon against plant maladies.

Micro-location

Often plots, especially those that are large or are located on uneven ground, are far from uniform. In a large plot with clay soil can contain a small area of sandy soil or even a rock outcrop. It is these areas of difference where varietal placement comes to the fore. If the difference are severe enough, these merit, as explained in Figure 7.1, another crop species.

Macro-location

Farms can be large or, because of topography and an irregular area, contain soil and site diverse plots. This is the basis for macro-location. Again, each plot is paired with one or more crops based on which are best suited for the location, only in this case the locations are well apart.

There is more to this. This comes in the form of exotic, non-native species that can strip a site of growth potential. In the wrong location, the tree eucalyptus is not a good choice.[8] The species is liked because it provides, on even water poor sites, strong, straight, and fast growing woody stems. Faulted for its aggressive water appropriation, this can, if grown in large groups or numbers, produce an adverse outcome with regards to localized water dynamics.

With this and like species, placement is critical. It may be best to place such a plant on the top of hills where the unsavory plant characteristics to not impact the better lands. As a warning, no species should be introduced unless the control criteria, such as the ability to contain unwanted spread, are unerring.

ASSOCIATED AGROTECHNOLOGIES

To derive an appropriate agrotechnology, varietal, and locational needs must transcend the simple mixing of varieties or species. As the planning is more sophisticated, more knowledge is needed. This relies upon an intimate knowledge of the genotypes available and how these are best utilized.

There are few associated agrotechnologies that clear fall under these heading. Two that have been identified are elevational and scattering.

Elevation

A common locational criteria is elevation. Near or at the bottom of slopes, soils are generally richer and more moist, at the top, nutritionally weaker, drier soils are often encountered. The elevational gradient represents more than altitude, these differences can be exploited to position an appropriate crop or variety. Hillside contour strips can incorporate an elevational agrotechnology to advantage.

Scattering

The agrotechnology most associated with location is scattering. In some regions, a farmer may own or use land across wide area, each plot separated from another. Farms of this type may be more common in topographically and climatically challenged regions.

Two thoughts may underlie scattering. The first holds that there is not enough high-quality land for all. As more enter, the farmers may divide the good as well as the lesser quality lands. In doing so, the plots of the newly created farms may become non-continuous.

Outside of the equality aspect, there is risk. Having plots on differing areas means that not all are subject to the same level of threat. When climate threatens, some plots will do better than others. Also, crop-harmful bugs will be forced to travel far to find their preferred meal. In league with fallows and other insect reducing measures, this can mean arriving too late to be a serious factor. To counter risk, it holds that scattering and location go hand-in-hand.

There are examples of this; the Incas divided and scattered land, insuring crop yields in bad years may have been the overriding motive. This is also demonstrated in the construction of irrigation systems, hillside terraces, and produce storehouses, all which are anti-risk. The Incas, as well as the current occupants, have multiple varieties, mainly of potatoes, enough to fully exploit a scattering strategy.

A second example is found in East Africa. Although the history is less certain, the motives may be similar. Here again, there exists a proliferation of varieties, principally of maize, to take full advantage of widely placed plots.[9]

ENDNOTES

1. The statement on the green revolution and national objectives are from Polak (2005), more general accounts, for and against, are in Stakman *et al.* (1967) and Shiva (1991).

2. Breeding plants for low-input systems is a suggestion by Phillips and Wolfe (2005), Atlin and Frey (1989), and Smith and Francis (1986).

3. The argument has been made that reliance upon a single staple crop is not good for human health (Pollan, 2003).

4. The choices for wheat, at over 600 varieties, is from Percival (1922). There are numerous other references that show, for each specific crop, varieties abound. The number of varieties available as seed stock is far smaller.

5. A sampling of studies on varietal selection criterion for non-commercial farmers are Gold *et al.* (2002), Benin *et al.* (2004), Wilson *et al.* (2003), and Lacy *et al.* (2006). A general discussion is in Goff and Salmeron (2004).

6. Among those finding a locational emphasis are Castelán-Ortega *et al.* (2003) and Samita *et al.* (2004).

7. This is a topic with immense complexities. A good narrative on exotic pests chasing exotic crops is by Jones (2006). It should be mentioned that introducing new plants does carry risks. Insect and diseases do hitchhike to infect native and domestic plant species, or these can become weeds. As such, movement can be a treacherous undertaking.

8. Williams (2002), among others, fault eucalyptus for non-environmental compatibility outside its native Australia.

9. Scattering in Peru is discussed by Zimmerer (1999) and, in East Africa, mentioned by den Biggelaar (1996).

8 Land Modifications

For many practitioners, the quest for higher and/ or assured yields are a strong justification for a lay-of-the-land change. The moving of earth can be an expensive investment, in time and energy, but, in the longer run, it secures continued and risk-reduced cultivation.

APPLICATIONS

The land modification vector is often found where crops will not grow or their growth is accompanied by a high risk for crop failure. In these cases, landusers must take measures to secure viable yields.[1] This often occurs in regions where nighttime temperatures plunge and/ or water shortfalls are a concern. In dry highlands, both can occur.[2]

Some societies prefer to raise crops that are climatically inappropriate. Terraces, paddies, hill-contouring ditches, and the like, all address this problem, either as a counter to natural climatic forces and/ or in topographically unsuitable terrain. Although the problem set is relatively small, the solution range is impressively large, there being plenty of agrotechnologies upon which to draw.

UNDERLYING BIODYNAMICS

The governing dynamics can include protection against rainfall, heat and/ or cold, drought and/ or wind protection, or other extremes in area weather patterns. The three things that land modifications are expected to do are:

- moderate extremes in temperate,
- increase the amount of crop-available water, or
- decrease the destructive impact of inordinately heavy precipitation.

Water Dynamics

Rainfall can be satisfactory and well timed or, as a countervailing force, heavy and/ or infrequent. Heavy or infrequent rains bring on problems must be addressed, both in protection from and effective utilization of that which is available. If shortfalls are severe, water can be brought from afar, irrigation channels are part of land modification as are paddies and other techniques of effective utilization.

Runoff

Water, as rain, is especially dangerous if the resulting runoff is of high volume, carving a destructive path. In addition to the loss of water as an essential resource, this can carry away soils and their contained nutrients. The counter is to allow the water to linger in-site, percolating into the soil, traveling belowground to the intended destination (hopefully the roots of crops).

There are advantages in replacing unconstrained surface runoff with slower belowground movement. Water, as a limiting resources, is available for a longer period if flowing underground at a leisurely pace. Subsurface water does not carry off soil particles. Also, any dissolved nutrients can be recaptured by plants with roots in the slower moving belowground current.

Surface runoff is first forestalled through a ground cover. Vegetation, biomass, or an overlay of stones all help. Water absorbing biomass is best, especially when rainfalls are infrequent. Stones store very little water but protect the soil from drying and do help with infiltration. Biomass and/ or stones requires that over 50% on the land be uniformly covered for a measurable effect.

Another lines of defense include ditches, bunds, and hedgerows singly or in combination. These are placed perpendicular to water flow, their spacing and size such that water from above-average rainfalls is pooled for future absorption. Well-spaced and well-placed barriers can increase water infiltration by 75%.[3]

Drought

The opposite of drought is flooding. Many of the counters for runoff also work against drought.

Keeping water on-site with ground-covering biomass, ditches, and/ or bunds are a foremost anti-drought defense. The idea being that pooled water, that constrained from flowing, is eventually absorbed and held on-site. Another counter to scarce water lies in reducing plant transpiration. For this, windbreaks or other wind-halting vegetation are especially effective.

Flooding

To counter excessive water, two strategies are possible: (1) keep the water on-site and (2) install anti-flood defenses. The first of these, often as ditches or bunds, are the initial runoff defenses. If region wide, these can do much to reduce what could be high downstream water levels.

When runoff defenses are not enough or seriously degraded uplands can no longer retain water in sufficient volume, lowland plots can be protected against the ravages of a flood. Safeguards include blockages in streams where normally dry channels hold soils, not always in-site, but at least in close proximity.

Temperatures

Extremes in the ambient temperature do influence yields. Cold, as well as heat, damages or kills crops. Where a problem, farmers have found ways to overcome.

Standing water and/ or stones moderate temperatures in the immediate vicinity. Standing water between mounds or crops grown in near rock piles are ward off spikes or plunges in temperate.

Another anti-temperate strategy is elevation. Cold descends affording protection to crops elevated above the surrounding area. This can be a field on a hillside or crops on mounds. Although mounds can be comparatively small, e.g., less than ½ m in height, this does keep temperatures a few degrees higher, often enough to avert major damage.

The presence of decaying vegetation, near crops, can also prevent cold-related damage. As vegetation decays, heat is released. Although the influence is not strong, this can again increase on-location temperatures by a few degrees, enough to reduce or eliminate harm.

Relevant Guidelines

One appropriate land modification can allow the growth of crops in normally unfavorable or chancy climes. More than one modification may be needed to make this fully possible.

In high altitude, hilly regions, those where the nighttime temperate can suddenly dip, heat-holding stone terraces offer a solution. In high altitude, flat regions, brief period of low temperatures can be countered by crop mounds or rocks placed around crops. If the situation is severe, mounds can be surrounded by standing, heat-retaining water.

Brief intervals of high rainfall followed by long, dry periods can tax crops. In addition to the control of surface runoff, there is the problem of destructive water flows along streams. An apt series of countermeasures (more than one are often required) will keep more water on-site, watering crops between rains.

Where the rains, not the temperatures, are the problem then a good ground cover with some form of in-plot infiltration, i.e., cross-hillside mounds, may suffice. Where both occur, the counter option may be a series of smaller modifications, e.g., temperate-holding rocks and moisture-collecting ditches near each plant. The agriculturally ruinous combination of steep, unstable slopes, heavy rains, and high and/ or low temperatures may require a significant intercession, such as stone terraces.

ECONOMIC MEASURES

Soil erosion, whether constant or the result of rainfall extremes, can be estimated before and measured after this occurs. This can be converted into a rough yield loss translatable into income units. The construction costs for various land modification, e.g., terraces, mounds, or ditches, are also estimable.

The next step in determining the feasibility of major construction is to compare the income loss against the cost of the countermeasure. This is done through conventional accounting. The standard approach is net present value (NPV). The criteria are that the NPV, a single derived figure, be positive.

As an example, take the case where soil losses reduce yields 20% as measured against the pervious year. Holding the value of the crop constant, this analysis is:

Year	*Expected revenue*	*Yearly revenue loss*
0	$2,000	0
1	$1600	$400
2	$1280	$720
3	$1024	$976
4	$819	$1181
.	.	.
.	.	.
19	$37	$1963
20	$34	$1959

With a normal discount, i.e., 4%, and over a 20-year period, NPV of the yearly revenue loss would be $18,863.[4] The conclusion, based on the break-even, that no more than $18,863 be spent for terraces, bunds, ditches, or other countermeasures. If the cost of a stone terrace is greater than the $18,863 figure, a progressive vegetative terrace, being a cheaper alternative, would be looked at. This type of analysis is often mandated by lending institutions, commercial farms, and absentee landowners.

With conventional NPV, highly successful land modifications have, in the pre-implementation phase, been shown not to be best option. This is solely analytical issue. Once in place, these often achieve a very profitable run. The problem is the early period bias of NPV. The higher the discount rate, the greater the bias.[5]

The next step is analysis from the point of view of a subsistence farmer, one that insists on food and income into the foreseeable future. It should be noted that, at a 0% discount rate, the value of terrace construction should not exceed $32,000 (i.e., the sum of yearly losses). This represents a willingness to spend more, $32,000 vs. $18,863, on revenue-saving construction. In not willing to take yearly losses, subsistence farmers are more willing to make the investment.

A few societies take a longer view and, through this, the subsequent analysis merges with the cultural/ societal interface. Farmers in long-view societies want an assured food supply for themselves and their offspring. With a 100-year timeframe, the break-even value of terraces or other countermeasures, without a discount rate, is roughly $200,000. This represents the amount a far-thinking culture, utilizing NPV, is willing to invest.

An actual NPV calculation falls by the wayside in subsistence or other cultures, but the idea remains, that of investing more where the cognitive timeframe is far-reaching. The notion being that, if labor and materials are available and the need is real, the project will find favor irrespective of the NPV.[6]

This informal analysis is borne out in the finely hewed, stone terraces of the ancient Incas. They had the time, the materials, and concluded that good, long-lasting, and costly construction was in their best interest. Their descendants do not have the exact same perspective, terrace are still of stone, but far less finely made.

Land modifications are often more pressing as they provide a risk-adverse cropping experience. For this, the risk index (RI) is more telling. This already incorporates a timeline. Stone terraces, in moderating temperatures, are less crop risky. It is for good reason that these are encountered where this danger looms.

Additionally, one must add in the intangibles and non-quantifiable benefits, e.g., terraces are a handy place to work as compared to a steep slope. Here again, if judged a good investment using a combination of NPV, risk analysis, intangibles, and cultural/ societal values, high-priced land modifications should be seriously considered.

DESIGN VARIABLES

Land modification structures can be looked in terms of their dimensions, e.g., hole depth, terrace height, or their interstructure spacing. These can also be viewed by through auxiliary uses and their number and placement across the farm landscape.

There are two design possibilities: (1) a larger interstructure distance, but with deeper holes, higher ridges, or taller terraces or (2) shallower holes, smaller bunds, or less tall terraces, but with less expanse between the structures. As for the actual dimensions, there are published guidelines,[7] but observation is an equally effective guide, e.g., Infiltration structures should be able to contain higher than normal rainfall without spillage.

ASSOCIATED AGROTECHNOLOGIES

As mentioned, there are ample land modification possibilities. With so many available, there is the opportunity to find the best possible option for any given situation. The options are often first evaluated as to their ecological effectiveness, the implementation cost, although important, may be a secondary decision. Land modifications are mostly revenue-oriented additions.

The agrotechnology list is as follows:

- Absorption zones/ micro-catchments
- Infiltration contours
- Terraces
  - Stone
  - Earthen
  - Progressive
- Paddies
- Ponds
- Gabons
- Waterbreaks
- Cajetes
- Water channels
- Mounds and beds
- Stone clusters

Absorption Holes/ Infiltration Ditches

Where rainfall is short or infrequent, it is better that the water, instead of running off quickly with accompanying erosion danger, permeate the soil. Small diversion ditches, installed slightly uphill of each plant, will divert scarce rainfall. Next to each plant is a small hole to hold this water in place. This design serves well with individual trees or with small clusters of mixed species.

When there are long periods between rains, another option can be presented. Adsorption zones are holes filled with water-retaining biomass, e.g., decaying leaves are one way to hold moisture. In windy, arid regions, the trees should be planted in the hole, affording some protection from drying.

In arid India, tree survival increased from 50% to 90% when individual capture ditches were installed. On the negative side, the cost of planting increased 20–30%. In the Sahel of Africa, the holes were filled with manure, retaining more moisture and, as a mini-environmental setting, attracting dung-eating termites. The latter further boosted soil fertility.[8]

Infiltration Contours

Small catchments, those that direct water to individual plants, serve well when tree are the recipient. Crop plots require a different design. For this, shallow ditches or bunds, placed along the land contours, will slow surface water (as in Photo 8.1). As insurance again spillage, segmented ditches (with cross-ditch barriers at 3–5 m intervals) will prevent water loss when part of the structure is breached or when leveling is not precise.

PHOTO 8.1. A hillside showing contour ditch–bund structures designed to slow and capture flowing water.

Studies show that well-spaced ditches or ridges can seriously reduce erosion. Research has shown soil losses 40–60% less when contour plowed fields have spaced infiltration structures, when these were covered with vegetation, erosion was reduced 90%.[9]

Terraces

Another approach to hillside management is to level or reduce the slope. Expensive to install, terraces are found where inclinations are generally too steep for cropping activity.

Within terrace category, there are sub-designs. The goal of a flat terrace is to convert a slope into a series of flat, contour strips. This type is the easiest to use, the most expensive to construct (Figure 8.1).

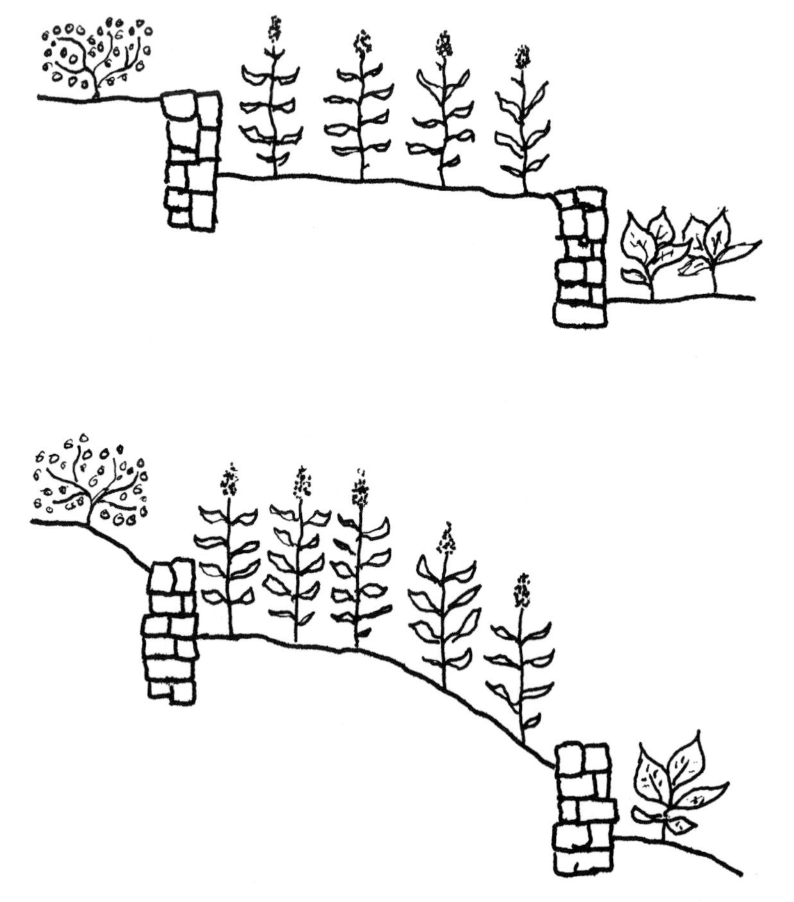

FIGURE 8.1. Stone terraced hillsides in cross-sections: the top shows a level terrace and the bottom has a reduced slope type.

Some may decide that, instead of a flat agricultural surface, it is just as convenient to merely reduce the slope. In some cases, the terrace is still along the contour, but the land slopes slightly from front to back of the resulting crop strip.

Terraces do not have to contour the slope. For this, there is the ramp type, called this because, in many designs, it is possible to walk from one terrace to another without climbing over the actual terrace structure. The unique feature is that the slope has two planes: (1) from back to front and (2) from side to side.

For fully constructed terraces, farmers generally work, for convenience, from the bottom of a hill. The other option, still bottom up, is a progressive terrace where the land is leveled, not in one-time period, but over a series of seasons. Both types, single period or progressive terraces can be stone faced or earthen with a vegetation covered slope.[10]

Stone

The steepest slopes may require stone facing. These can be progressive, adding more rock on the top of the structure as the soil accumulates, or constructed in one short span of time. These can be fully flat along the contour, have an inclined agricultural surface, or be of a ramp type.

There are other designs, the box or coral terrace has small plots with stonewalls on four sides. The upper and lower faces hold a flat or sloping cropping surface. The side or down-slope walls contain grazing animals or protect the small crop parcels from grazing.

Stone terraces, being expensive, must provide multiple gains. Soil erosion should be eliminated, in arid regions, more water should be retained on site, and the terrace stonework should moderate extreme hot and cold temperature.

Earthen

Instead of stone, an earthen structure can provide either a flat contour, sloping contour, or ramp design. To keep the terrace face from eroding, it must be planted with protecting vegetation. Grass-cover terraces may be the most common form.[11]

There are different notions as to the type of terrace. Grass or tree-supported structures do not have heat-holding capacity, but still allow cold air to descend, away from the elevated and frost-susceptible crops. For most designs, the terrace itself need not be flat. The most common exception is for hillside rice paddies, here a horizontal area, one capable of holding water, is an absolute precondition.

Progressive

Rather than building terraces one-by-one up slope, the alternative is progressive establishment, this is where each terrace is formed over time. For this, lines of trees or other sturdy vegetation are established. The face of a hill is plowed, the eroding soil is caught against the tree lines. With proper management, the trees are pruned and the pruned branches piled along the upper side of the tree lines. In this way, more soil is caught. Over time, a progressive contour terrace will emerge.

As mentioned, this can also be accomplished with a stone construct. These tend to be less utilized in a progressive form.

With a progressive terrace, labor and costs are spread over time, making this a less daunting economic investment. This has an added advantage in that nature, through regulated erosion, aids in the construction. The disadvantage is that it takes many years for the terrace to materialize, an aggregate of 6 years has been reported.[12]

Paddies

Pools of standing water with little depth, i.e., paddies, are commonly associated with rice cultivation. Other crops besides rice (*Oryza* spp.) are grown in this swamp-like environment; examples include cranberries and wild rice (*Zizania aquatica*).

Out of necessity, paddies need a flat surface. This can be swamps flattened for agriculture, level fields with bunds to hold the standing water, or, in the most dramatic form, flat, hillside terraces which are modified to form shallow ponds.

As major investment, these are expected to produce high-valued crops. Use is generally intense enough so that irrigation is also economically justifiable. If not utilized for a water-loving species, paddies are a prime location for other high-valued crops.

Ponds

Ponds are less a medium for crop growth, more an auxiliary, multiuse structure. As such, these have many ecological and economic roles. Included is water storage for irrigation, animals, or household use. Ponds can also be fish source, for income or food, and a place where some species of insect-eating insects live and breed (Photo 8.2).

Where rainfall is heavy and infrequent, seasonal ponds find use. These are low areas near streams or normally dry watercourses. When water is flowing, some of this is channeled to these low areas to form temporary ponds. The reason may be to provide water for animals or crops well into dry seasons.

PHOTO 8.2. A general-purpose farm pond.

Gabons

To increase in-site, in-soil water, solid, large barriers can be placed across active streams or normally dry channels. These gabons, often as rock-filled wire baskets, have a twofold purpose. The first is as an anti-flooding measure, helping to trap and retain water-carried soils during periods of high water runoff. The second purpose is as an anti-drought measure. The movement of water is slowed, allowing more water to seep into the ground.

These are most effective when first construct in the upper reaches of watersheds. The installation generally progresses from the higher to the lower elevations. The purpose of this ordering is lower the force of the moving water and avoid overwhelming those in the lower elevations before the upper are in place. The economic gains come in keeping erosion in check and in protecting lower elevation plots. A secondary benefit will be in increased moisture availability, again for the bottomland plots, those below the gabons.

Waterbreaks

As with gabons, waterbreaks impede water flows. Instead of blocking active or less active watercourses, waterbreaks cross plots that are inundated during floods. The intent is to stop soils being washed away and, by slowing water, cause dirt in the water to be deposited.

In the best of circumstances, the hope is to also replenish nutrients. As waterbreaks can play other ecological roles, e.g., serve a part of system of windbreaks or be a living fence line. If the case, cost can be amortized against the various tasks undertaken.

Cajetes

Rather than having level, contour following ditches or mounds, the option exists to utilize an irregular sequence of deep holes. This are mostly found where cropping areas are rocky or the ground highly uneven, making it difficult to install cross-slope infiltration structures (see Photo 8.3).

To make these more effective, diversion ditches can be installed to channel the water flow into the hole. Since the object being to capture the rainfall without runoff loss, this is an informal guide (that of collecting all the water without runoff) which dictates the size and interhole distance.

There are two uses for cajetes. The first is for orchards or other treecrop systems located on rocky or irregular hillsides. The second is where steep, rocky hillsides surround flat, crop suitable bottomland where the runoff captured is intended to boost yields for the downslope plots.

PHOTO 8.3. One of many cajetes, large hillside holes dug to capture flowing water that are utilized when the terrain is too irregular for contour ditches.

Water Channels

Found on many irrigated landscapes, water channels convey water across the farm landscape. These come in many forms, the better designs are not merely a means to transport water, but are agricultural relevant and part of the ecology of a farm.

The vegetation, mostly trees, along the banks of channels can be a windbreak or a movement habitat for predator insects. The banks themselves, being well watered, can be long narrow strip devoted to moisture-loving crops.

The economics are simple, the end use of the water should justify the installation and maintenance expense. Any secondary gains, as briefly listed above, only add to their utility.

Mounds and Beds

In many highland regions, mounds counter brief periods of low temperatures. The principles that underwrite these are a slight elevation-related temperature difference, the mound is surrounded by heat-retaining water, and/ or decaying vegetation inside the mound that releases heat.

An elevation-related temperate difference may seem small, e.g., a 0.5 m high mound can have a 2°C temperature difference between the top and bottom, often enough to prevent damage to a plant.

Regions where this difference is important include the highlands of New Guinea and Bolivia. In the past, Native Americans once employed mounds for growing maize and, in doing so, safeguarded their crop from early season frosts.[13] Where utilized, these are regarded as a necessary cost to insure against crop failure.

Stone Clusters

Although rare, some societies grow young trees or large annual plants between medium sized stones or within donut-shaped rock clusters. Generally, the placement of rocks, from 5 to 25 cm in diameter, around the base of plants or across and under perennials, will moderate the air temperature, make the soil surface receptive to water infiltration, trap air-borne nutrients, help fend off weeds, and offer a situation discouraging to those small animals that eat crops or the plants themselves.[14]

Clusters can also be general ground cover or in rows, not just piles around individual plant. This would be for perennial species, the purpose is temperate moderation.

Rows of stones can also be placed along the upper side of contour mounds. This might be of dual purpose use: (a) temperature effect and (b) to speed water infiltration.[15]

As with mounds, there is an additional consideration, fields littered with stones are less suitable for mechanized agriculture. Where tractors operate, other temperature counters, e.g., treerow alley systems, might be used.

ENDNOTES

1. Rainfall catchments and distribution systems, widespread in past societies throughout the worldwide, is proof as to their effectiveness. Currently, these are underused and under recommended and, if better promoted, could have large regional impacts (Coghlan, 2006).

2. The problems of temperate and water are worldwide, a sample from the literature shows this in Bolivia (Zimmerer, 1999), Pakistan (MacDonald, 1998), New Guinea (Waddell, 1975), China (Wang, 1994), and South America (Caramori *et al.*, 1996).

3. The 75% infiltration figure is from Hulugalle and Ndi (1993).

4. Because NPV is a well-recognized technique, the mathematical details are not presented.

5. Price (1995) discusses the NPV as a viable determinate.

6. Additionally, very large projects, such as irrigation systems, may require strong governmental or similar control, able to mobilize dedicated labor.

7. Published guidelines for contour ditch spacings are in Weber (1986).

8. Dung-filled capture holes is from Glausiusz (2003), use in India is from Gupta *et al.* (2000). Other references for absorption holes are Nyakanda *et al.* (1998) and McIntyre (1999).

9. The reductions in erosion from spaced ditches are from ICRAF (1993).

10. A study of terrace layout can be found in Schulte (1996). Wojtkowski (2006a) gives a brief overview of stone vs. grass or tree-based terraces.

11. Uses of grass-covered terraces are in the Philippines (Stark *et al.*, 2000) and in Honduras (Walle and Sims, 1998).

12. The 6-year figure for terrace formulation is from Banda *et al.* (1994).

13. Temperate and moisture are not the only reasons for mounds. Siame (2006) reports mounds, made from decaying organic material, implemented for their enhanced soil fertility.

14. The use of stone clusters is presented in Fish (2000), Gonzáles (2001, p. 142), and Pearce (2006) with ecological reasoning in Steenbergh and Lowe (1969).

15. Rocks atop hillside mounds are described by CEPIA (1986), no reason is given.

9 Cross-Plot Influences

Individual agroecosystems can be ecologically and economically autonomous, however, things are seldom this simple. Plots do influence, sometimes to a high degree, productivity in adjoining crop systems. This can be so pronounced that knowledgeable farmers will install ecologically beneficial systems next to those requiring such help. This is the basis of the cross-plot or single-effect landscape vector.

This can, and often does, go beyond one-on-one relationships. An entire farm, or even a region, can be agroecologically integrated so that cross-effects are many and pervasive. The unfortunate part is that practical development lags theory.

As a plot vector, the larger landscape is not the focus, the individual recipient agroecosystem is. The advantage of the narrow view is that complexity is limited to one-on-one, cross-plot influences.

APPLICATIONS

Economists often talk about the need to diversify as a means to protect against crop failure. The notion being that, if one plot succumbs to a climatic fluctuation, another might survive in a better protected dissimilar crop or in a different location. Growing different crops is also a safeguard against price change; if the selling price for one-crop drops that of another might rise. There are labor gains as unlike crops may require tending at varying times, this spreads labor across a wide time period making better use of what is available.

Given the above, there is ample reason for having a mix of agroecosystem types, the more diverse, the better. For maximum variability, farms can go beyond agriculture to incorporate forestry or agroforestry ecosystems. This expands the opportunities for cross-plot ecological gains.

Cross-effects should be pre-planned. If one ecosystem can influence those nearby, why not purposely position non-productive systems to accentuate and spread positive influences. These non-cropping ecosystems, those that only offer facilitative services to neighboring crop plots, are called auxiliary ecosystems.

In use, auxiliary systems have a major advantage. When shorn of the need for a profitable output, these can be totally dedicated to the ecological tasks at hand and very effective at providing cross-plot facilitatory services. This is not a requirement, productive ecosystems, those with strong ecological properties, can and do serve well as ecologically friendly neighbor.

UNDERLYING BIODYNAMICS

The underlying dynamic for the landscape vector lie in the myriad of cross-influences that can be manipulated. Winds, water, bugs, birds, bats, large animals, and plant diseases can all freely cross from one plot to another. Governing or exploiting these is part of an ecologically active landscape.

The common influences regulated through a cross-plot landscape are:

- spread of herbivore insects,
- transmission of plant diseases,
- flow of water across the soil surface,
- prevailing and storm generated winds,
- high and low extremes in temperature,
- movement of large, crop-damaging animals.

In controlling the above, there are dangers in aggravating a landscape, i.e., increasing some productively negative cross-plot effects. These include:

- increasing the force of winds (the wind tunnel effect);
- providing a habitat or refuge for crop-eating birds and destructive small animals.

Agroecosystem Properties

Behind the landscape vector is the notion that all ecosystems have defined properties. An agroecosystem may be prone to or resist water erosion; alternatively, plant-eating insects may find a crop type to their liking. The list of such properties, the desirable agroecosystem properties (DAPs), is long and can be ordered, meaningful to inconsequential, strong to weak.

For live fencing, examples of DAPs might include:

- serving as a windbreak;
- resisting wind and water erosion;
- harboring beneficial insects and, most important;
- obstructing the movement of large animals.

Although a specialized auxiliary system, live fences are best placed near productive systems that need these crossover properties. The placement of agrodiverse (productive) systems is an attempt to exploit whatever DAPs a neighboring system might have, e.g., ecosystems that are good at harboring predator insects can be placed next to those that lack, and require, these insects.

Interface and Spillover Theory

The more diverse ecosystems are in contact, the greater the opportunity, with some exceptions, for the desired cross-effects. Looking at one aspect, a landscape can be maximized around (a) the amount of interplot interface, (b) the desirable amplitude and store of the ecological effects in adjoining ecosystems, (c) penetration of the spillover, (d) minimum profitable cropping area, (e) the total area, and (f) host of other, influence-specific variables.

With insects as the dynamic, the spillover has the predator types penetrating into a insect-threatened cropping area. The affect-specific variables may include matching the life cycles of insects in question and the presence of winds that can help disperse the predator types.[1]

The idea behind this is that there exists, in abstraction, a perfect revenue line. This is the straight line in Figure 9.1. This establishes a direct relationship between percent of area planted and the yield and subsequent revenue. The assumption is that these crops face no yield reducing menaces.

Of course, this is not what happens. Crops are threatened and losses occur. If an area is fully planted and unprotected, insects will eat a high percentage of the crop. Greater yields can occur if some of the area is a breeding ground for predator insects as, e.g., a productively inert, internal auxiliary system. These predators then spillover onto the crop, resulting in relatively risk-free yields (Figure 9.1, Y_b instead of a probable Y_a).

A number of factors can further enhance the spillover effectiveness. More interface greatly helps, e.g., instead of an auxiliary system as a low-interface, square-shaped plot placed next to crop plots, these can be a high-interface, internal strip pattern. If the strips are too small,

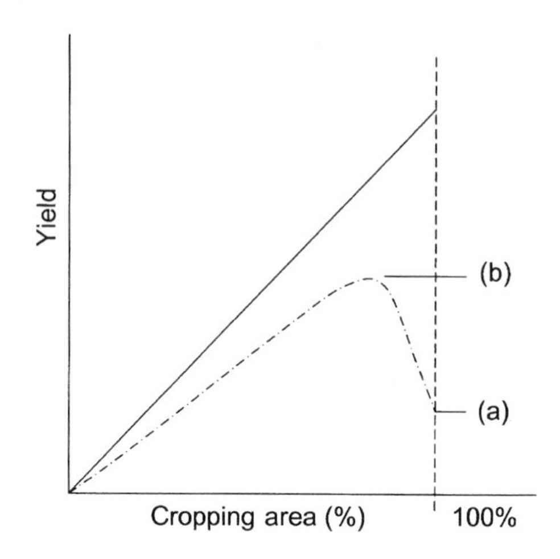

FIGURE 9.1. The theory behind anti-insect or other in-plot threat containing measures. Upper line relates plot area utilized, without threat, to output. With risk, rather than chance low after-event yield (point a), some of the plot area is utilized for threat containment and more confident yields (point b).

these may not be a good breeding ground, a good safe haven for predator insects, and poor at other risk-reducing tasks, e.g., slowing and absorbing surface water. If the strips are too wide, the reduced cultivated area also reduces potential revenue. As shown in Figure 9.1, there will be an optimal.

There are means to intensify the strips of natural vegetation that bisect crop plots. When herbivore insects are the problem, strips can be augmented with an insect-attracting plant species. The intent is to lure harmful bugs away from the crop to a zone of greater peril (through a decoy species) or to lure predator types.

There are many examples of lure-based agroecology. To attract harmful insects, napier grass tempts the maize stemborer. As a draw for specific predators, vetch strips can pull in leatherwings that eat cucumber beetles, dandelions lure ladybugs which feast on aphids.

There is the possibility of a push–pull arrangement. For this, repellents plants in the center of crop strip push while an attractant species in the center of the predator strip pull. The effect can be more crop area, less natural area. Expanding upon an already mentioned example, the plant desmodium repels the maize stemborer whereas napier grass attracts these insects.

The same conditions, illustrated in Figure 9.1, apply to other control scenarios. As with the control of crop-damaging bugs, more interface can help, e.g., to reduce soil–wind erosion, to control temperate extremes. The actual yield curves will vary depending on how the forces of nature are employed and, thus, will vary somewhat. Despite this, the theory is sound and this is reenforced in the many applications found worldwide.

Cost Control

When cross-plot influences involve the relative placement of two productive systems, the economic orientation effect on either should remain unchanged. If auxiliary systems usurp cropping area and reduce the revenue potential, this is clearly cost orientation. Auxiliary systems, because they can do so much toward crop beneficial ecology, are one of the more powerful agroecological tools.

Relevant Guidelines

For the landscape vector, one needs cross-fortification, i.e., the strengths of one ecosystem complement the weaknesses of another. Key is amount of interplot interface and the potency of the spillover. The other factor is to accomplish as much as possible with a few auxiliary systems. Strips of productively inert can, encourage large populations of predator insects, pull in harmful insects, function as anti-erosion infiltration contours, and undertake other useful tasks.

ECONOMIC MEASURES

The economic measures of the landscape can be based on financial criteria and/ or be made along more ecologically diagnostic lines. For the latter, the land equivalent ratio (LER), a standard in agroecology, can be expanded, providing the same insight into the underlying cross-plot mechanisms.

Multi-plot LER

The application of the multi-plot LER (MLER) can be illustrated when a plot is subdivided into a productive section and an accompanying non-productive auxiliary system:

$$\text{MLER} = a_1(\text{LER})_{a_1} + a_2(\text{LER})_{a_2} \tag{9.1}$$

where area a is equal to $a_1 + a_2$, expressed as proportions, these sum to one. The LER determination for a_1 is based on the primary crop yields obtainable from the original unreduced plot, i.e., plot a.

Expressed numerically, an example is

$$\text{MLER} = 0.8(1.50)_{a_1} + 0.2(0)_{a_2} = 1.20 \tag{9.2}$$

For this, 20% of a 1 hectare plot is appropriated for non-productive facilitative purpose, hence the 0.8 and 0.2 land breakdown. This subdivision, through the addition of an auxiliary systems, strip or otherwise, shows gains better than one, the taking of productive land for facilitative purpose patently justified.

DESIGN VARIABLES

For the landscape, there are a number of design variables, all involve the layout of the individual plots, both shape and size, and their relative location *via-á-vis* the recipient plots. The goal is to maximize the positive cross-plot effects, i.e., where one agroecosystem facilitates the neighboring crops.

Interfaces

When spillover effects are paramount, it helps that the interface between the facilitative-furnishing and the facilitative-receiving ecosystems are as large as possible. This most common layout has adjoining narrow strips, a second best is small, highly interactive blocks.

Buffer Species

At times, cross-influences between plots needs to be filtered or checked. Sunlight streaming into a block of trees from the side can cause an edge effect. This results in below-canopy weeds and, for the trees, excessive ground-level branching. The latter lowers the quality of the tree stems as salable logs.

Above and belowground, branches and surface roots from trees can extend over and under nearby crops, robbing these of essential resources. A blocking or buffered interface can be used. As mentioned in Chapter 5, barrier and boundary systems, these are tree or other perennial species with the needed desirable plant characteristics (DPCs), e.g., vertically shaped canopy and thick, deeply penetrating roots, so placed as to halt root intrusions. A single row of a buffer species can stop the competitive crossover effect from adjoining crops.

ASSOCIATED AGROTECHNOLOGIES

Within plots, facilitative plants offer biodynamic and economic services, similarly nearby plots do likewise. Landscapes often contain areas devoted to facilitative tasks, some accomplish the tasks as productive

PHOTO 9.1. Various landscape features including corridors/ barriers along with surrounding natural ecosystems. This photo is from Central America.

unit, others do so as devoted, non-productive, auxiliary ecosystems. Whether or not these are a source of weed seeds is an open question, much will depend on vegetative content of a strip or other feature.[2] As previously stated, these are mostly cost-oriented additions.

There are a number of acknowledged special purpose, auxiliary landscape features. A listing includes:

- Windbreaks
- Anti-insect barriers
- Habitats/ corridors (Photo 9.1)
- Riparian buffers
- Firebreaks
- Living fences

Windbreaks

The classic example of an interplot influence is the windbreak. These are standard features in many agricultural regions and vary much in form, location, and design.

This can be fairly simple spatial layout; rows of trees that cross the landscape at set intervals. This might also be a more complex system of shelterbelts, windbreaks, and temporary wind protections.

Shelterbelts are permanent, broad, multi-tree strips that cross the landscape at some distance. Because they contain, in cross-section, numerous trees, they can serve a useful role other than blocking prevailing winds. These can be for forestry, the harvest of marketable timber or firewood, or an agricultural addition, as a source for native berries, ornamental shrubs, and/ or medicinal herbs.

Windbreaks, in subdividing an area further, enhance the main purpose of the shelterbelt. These, being narrower, have less of a secondary purpose.

In some landscapes, less tall, seasonal windbreaks are added. These stand alone in wind sheltered plots or are part of a shelterbelt–windbreak system where strong, constant winds are typical. Seasonal windbreaks can be of a productive or non-productive species, common windbreak crops include sugarcane or Jerusalem artichoke.[3]

Practice, backed up by numerous scientific studies, has shown the windbreak to be an almost universal economic plus. With a well-designed cross-landscape system, yield increases of up to 70% have been reported. These studies apply to pastures as well as cropping systems.[4] There are additional gains that come in with an improved habitat for pollinating insects.[5]

These increases are on top of a few tangible, but incalculable, environmental pluses. The latter includes windbreak structures as a refuge for birds and small animals and as corridors for animal movement.

The major decision variables are the locations and the internal design of various wind structures. A well-functioning system of windbreaks, as shown by examples throughout the world, does produce favorable economics. Less noted are the associated environmental gains.

Anti-insects Breaks

An anti-insect barrier can be economically favorable landscape addition. Mostly, these are areas of grasses and/ or a mix of other herbaceous plants, areas well suited for predator insects. The chief gain lies in the spillover of predator insects onto adjoining croplands. This presupposes plenty of interface with the crop and, as a result, the designs are mostly in strip form.

Rather than homogeneous mix of vegetation, insect-attractant species, as well as vegetation favored by predator insects, can make these into a potent kill zone against herbivore insects. A few taller plants, placed in lines near the center, might help to halt plant-eating insects as they overfly the area, exposing them to the in-lurking dangers.

Also expected is some nutrient transfer. This can be through roots, spreading out from the break, cut during plowing, subsequently decaying in place.

The economic assessment comes in the reduction of costly insecticides or fungicides. The question being; does the inclusion of such break help the RVT or the $CER_{(RVT)}$ enough so that the cropping area, as facilitative non-productive strip, is more than compensated for through increased per plant productivity?

Habitats/ Corridors

Along the same lines as anti-insect strip or breaks are habitat areas. In size, placement and content, the purpose is to provide a refuge for and breeding place for those insects that eat other insects along with other insect-eating creatures and those that pollinate.[6]

There are other gains. Earthworms shelter and breed under these same areas, afterwards moving beneath crops.[7]

There is also the question of insuring spread. Small flora and fauna, including microorganisms do not move easily between plots of refuges, corridors help in this.[8] The organisms included are microbes, fungi, and earthworms and, aboveground, a host of more mobile, larger organisms, e.g., useful insects.

Corridors are not only roadways, but also habitats that constantly replenish beneficial, and some not-so-beneficial organisms, to adjoining plots. This is especially important when cultivation, going from one bareground phase to another, disrupts or ruins in-lying populations. Restocking is a necessity and corridors do this.

Riparian Buffers

There is a whole class of systems designed to keep moving and still waters clear and cool. Against this goal, unadvised agriculture, through soil and nutrient runoff, is a major cause of water contamination. Runoff in whatever form is an indication of poor agricultural practice as soil and nutrients are wanted by crops.

Of course, not all well-formulated agroecosystems are perfect, some do leak nutrients and riparian buffers are intended to plug these leaks. In general, buffers are located along the banks of wades, streams, rivers, and ponds.

There are agreed upon design parameters. The vegetative content should be thick, perennial, and growth active, capturing almost all the nutrients for as much of the year as possible. The latter is important as nutrients and soils can escape capture when buffer species are dormant.

The internal design of a riparian buffer is less of a factor than their location. In addition to an almost mandatory positioning ringing or bordering open water, they can be found in association with crop plots. The idea being that soils and nutrients are best intercepted as close to their source as possible and that a series of riparian defenses, within and around plots are preferable to a last-opportunity, stream-side shield.[9]

As a farm ecological structure, the environmental gains are real, but are intuitive and incalculable, the losses are monetary, calculable, and mostly negative. The latter occurs as the best quality land is taken out of production, i.e., those flat, well-watered, fertile areas along watercourses. This makes for tricky economics. If a calculation is required, a value must be put on clean, clear, cool water. However, most do without, the worth of riparian systems coming from an innate, not an actual valuation. Therefore, riparian systems are championed more for their intuitive, but incalculable pluses.

Firebreaks

Some crops and landscapes can be fire susceptible. End of season drying can put grains crops at risk. If blazes are relatively common, it might be wise to employ fire strips.

The most common anti-fire barrier is a plowed strip, clear of all vegetation. The width depends on the severity of a flare up. Alternatively, a planted strip is possible with species which do not burn, e.g., many cactus species fall into this category.

As a temporal alternative, the firebreak can be a crop-planted strip where, due to the timeframe of the crop, the strip has been harvested, is bare and free from vegetation when the fire danger is greatest.

The economics of a firebreak are mostly risk motivated and evaluated accordingly as these strip offer few other ecological benefits. Where crop insurance is available, these are readily dispensed with.

Living Fences

Barrier are part of many landscapes. These can be as fencing, with the express purpose of keeping people and animals in or out of farm fields. Dead fencing, with non-living posts, is ecologically inert and of little agroecological interest. As economic structures, these are evaluated the same as barns, chicken coops, and other farm structures.

The ecologically active live fencing can do more than just blocking the passage of animals. If so designed, these can have anti-insect properties and serve as interplot connecting corridors.[10] Living fences come in many forms, from live posts with wire, live posts with interwoven branches, multi-post, to thick-shrub designs. Figure 9.2 shows sampling designs.

Given that fences are the most universal of all farm features, their worth is not questioned. The economic issues are on which designs are best in any one given situation. Many choose dead posts because they are available for immediate use and require less yearly maintenance.

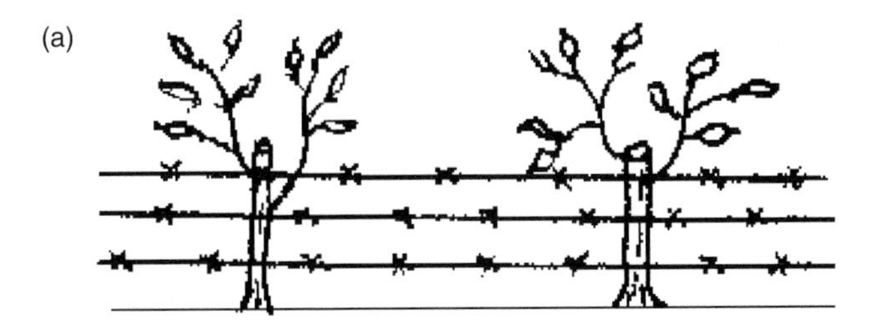

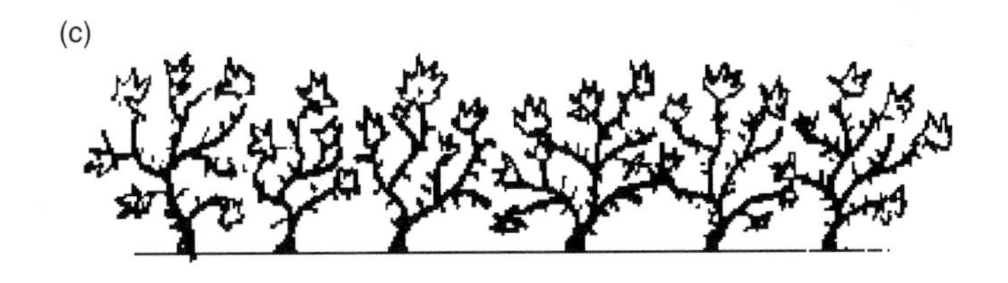

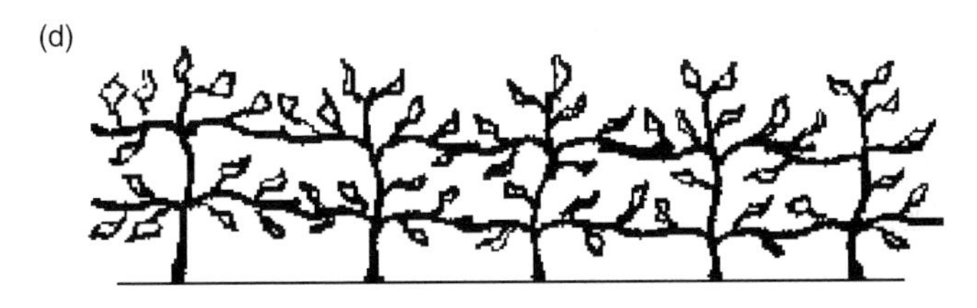

FIGURE 9.2. Four living fence types. From top to bottom: (a) wire on live posts, (b) a multi-post design, (c) a thorny or spiny hedge, and (d) pleached hedge.

The disadvantage is that they must be replaced at frequent intervals, at a large cost outlay, unless the posts are highly decay resistant.

The live versions take a long time to establish, must be protected from grazing during this period, and do take low levels of yearly inputs. Because of this, hybrid designs (e.g., thorny vines overlaying a dead fence) can, if the economics are favorable, offer the best of both worlds. Other variations are temporal, live fences are established while the dead version is in use, eventually one replacing the other.

The overall economics is not complex, but involve a lot of decision variables and a few tradeoffs.[11] The transition must be considered and it can be difficult to fully assess the pros and cons as well as the ecological gains for the various living designs. The decision is made simpler if fences multi-task i.e., the fence are the ecological centerpiece of a vegetative, predator-insect harboring, wind blocking, and animal obstructing strip. Multi-tasking can more than vindicate any additional costs of a living fence.

ENDNOTES

1. Questions on the value of insects (predator and/ or pollinating) has been discussed in macro terms, e.g., Berger (2006), less so when individual plots are involved.
2. The question on weeds from auxiliary systems is partially answered by Devlaeminck *et al.* (2005). The finding is that forested landscape features do not spread unwanted plants.
3. Details for shelterbelts and windbreaks are in Wojtkowski (2004, p. 151) and Caborn (1965).
4. The 60% to 70% yield increases are for maize and millet (Zhang Fend, 1996) lesser, but still acceptable gains in the 10% to 40% range, have also been reported (Brenner, 1996). Increases for grazing animals are in Prinsley (1992).
5. Windbreaks afford a favorable habitat pollinating insects, especially if flowering species are included in the windbreak. The most crop-favorable conditions may occur if the in-windbreak species do not flower at the same time as the crop species (Vaughan and Black, 2006a, b).
6. Among the references for agricultural habitats are Alomar *et al.* (2002), Vickery *et al.* (2002), Bäckman and Tianen (2002), Maudeley *et al.* (2002), Groppali (1993), Ma *et al.* (2002), and Altieri and Nicholls (2004, p. 131).
7. The case for earthworms in herbaceous strips is from Lagerlöf *et al.* (2002).
8. The benefits of corridors for moving various organisms is demonstrated in the works of Matlack (1994) and Peterken (1993) with discussion on the role and design by Laurance (2004). Effectiveness at the landscape level has been verified by Damschen *et al.* (2006).
9. The discussion of riparian systems continues unabridged in Chapter 14.
10. Live fences to ecologically connect plots are discussed by León and Harvey (2006).
11. A study of localized fence acceptance factors, including the prevalence of forage-seeking cattle and protection-requiring crops, is found in Ayuk (1997).

10 Ex-farm Inputs

The agricultural ideal is let nature provide to the degree that a farmer has only to harvest a crop. A second best scenario is to plant and harvest, with no other additional costs. Unfortunately, nature is not that kind, farmers must restock soils with nutrients and defend the crop from the full range of threats.

In aiding crops, outside or ex-farm inputs have advantages. For one, these avoid the complexities of biodiversity. Of course, these have drawbacks, economically foremost is the cost of purchase. As many arrive as manufactured, non-natural chemicals, there can be an environmental toll as exotic compounds can and do poison natural organisms. This does not only apply to chemicals formulated to kill, i.e., insecticides, rodenticides, fungicides, and herbicides, but also to synthetic fertilizers which, over the long term, can alter soil health in a not so positive way.[1] Under the do-no-harm admonition associated with agroecology, agrochemical use requires forethought and only under severe guidelines.

The chapter looks at how and when ex-farm inputs are used, not in a conventional-based monoculture, one totally supported by outside agrochemicals, but in an agroecological augmentation. For insecticides, herbicides, fertilizers, and fungicides, the concept here is a bit different than presented through the monocropping norm. For labor and irrigation, the applications follow time-tested principles. For other inputs, e.g., introduced predator insects, the attempt is to mimic what nature does.

APPLICATIONS

With a conventional approach, farmers wait until something happens and then apply a chemical response. Potent ex-farm inputs may seem, on the surface, the best immediate response. By digging deeper into the agroecological toolbox, it is possible to find a better course.

When plant-eating insects strike, a farmer can spray insecticides, go out and pick them off by hand, or introduce chickens or some other insect-eating fowl. Picking them by hand is not often viable, but chickens are a form of natural control, picking insects is what birds do naturally. The farmer can also purchase and interject predator insects. Other options include insect traps.

In suggesting a range of solutions, chemicals are not entirely eliminated as an agroecological tool. For centuries before the advent of synthetic chemical compounds, farmers found these in natural form. Broadly referred to here as home remedies, there are many benign, nature-safe possibilities.

Low on any application list are those harsh chemicals that are environmentally incompatible. As well as harming the good insects, these can be a risk to worker health, can destroy local fauna, contaminate land and water and, when poorly used, can cause insect populations to flourish after the initial impact has subsided.[2] If they must be employed, these are spot applied in very small amounts, not broadly sprayed, and then only as a last crop-saving defense.

UNDERLYING BIODYNAMICS

In taking full advantage of what nature has to offer, in term of substitutions, practitioners have the option of seeking multiple gains from safe, well-directed ex-farm inputs. These can be expected to:

- increase yields,
- decrease weeds,
- stop plant diseases,
- eliminate insects,
- arrest plant-eating rodents.

Yield Gains

The application of fertilizers and irrigation can be tracked through production functions. These describe a mathematical relation between the amount applied and the resulting yield levels.

The classic, and hypothetical, single-resource function is shown in Figure 10.1. Usually this is with the limiting resources, all other essential resources are assumed to be in abundance.

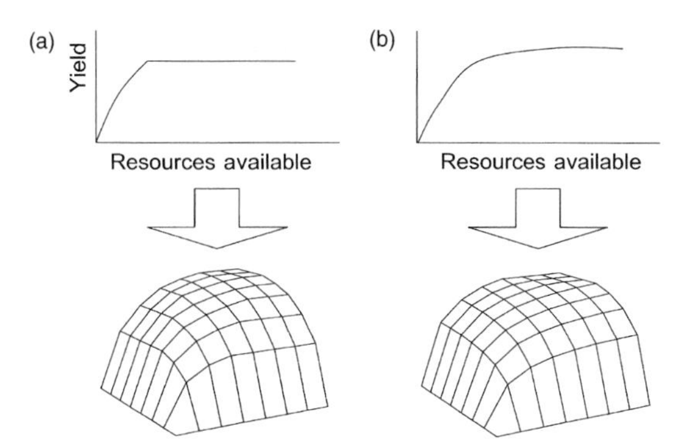

FIGURE 10.1. On the top are the single-resource functions under two hypotheses: (a) the von Liebig and (b) the Mitscherlich. Although theoretically different, when transposed into a multi-resource form (bottom), these become visually similar and have proven hard to statistically distinguish.

When more than one essential resources is involved, the picture becomes less clear. There are competing hypotheses on what happens. There are two major hypotheses: (1) the von Liebig and (2) the Mitscherlich. Basically, the von Liebig hypothesis assumes one, and only one, limiting essential resource. This limiting resource sets the yield level, irrespective of the amounts of other essential resources.[3]

In contrast, the Mitscherlich hypothesis, expressed in the Mitscherlich-Baule form, also has a single and limiting essential resource. Instead of being absolute, yields can exceed the limit set by the single resource if the other resources are in abundance. Graphically expressed, Figure 10.1 shows, in comparison, the single input von Liebig and Mitscherlich. Although chemical fertilizers may be the most common form, ex-farm nutrients also include cut-and-carry biomass, manures, or compost.

Risk Abatement

The risk reducing potential of ex-farm inputs goes unchallenged. If the crop is in danger of failing, applying that which is missing can remedy the situation. The philosophy is clear and the results are immediate. As these remedies can be deceptively simple, risk abatement through inputs has a strong following.

Climate

Water is often lacking and remains a critical input in arid regions. As an ex-farm inputs, the amount applied follows standard analysis

(Figure 10.1). Of course there are different forms of irrigation, some more efficient that others, each with an environmental presence. The pros and cons of irrigation are presented later this chapter.

Insects and Diseases

In agroecology, a number of ex-farm countermeasures for insect and disease outbreaks. If the in-place countermeasures prove weak and are overwhelmed, then an ex-farm input would be considered. It is far better that these are part of a pre-planned strategy, failure of the first-line defenses will trigger introduction of a benign ex-farm counter.

For the good, insect-eating insects can be purchased and applied. Ladybugs, which feast on aphids, are a possibility as are species of wasps which are less specific feeders. Ants and spiders also control various insect pests.[4]

As for chemical controls, gardening books list many benign home remedies. To name a few, milkweed seeds destroy nematodes and armyworms. Diatomaceous earth causes beetles to die from dehydration, its use as a food additive demonstrates safeness. As a substitute for fungicides, milk, sprayed as a dilute solution, reduces leaf mold.[5]

ECONOMIC MEASURES

The overriding issue may lie in the economics of substitution. As mentioned in the introductory chapter, there is the conventional, revenue-oriented approach that provides the full yield potential but through ample and expensive inputs. Revenue orientation greatly simplifies things.

In contrast, agroecology offers cost-oriented approaches that tolerate some yield loss, but uses fewer and less costly inputs. Within an agroecological context, revenue orientation is also possible. Once the foremost issue of economic orientation is resolved, there are many questions on the options and outcomes.

Some issues arise in the amount of ex-farm inputs applied, specifically application in economically optimal quantities. Key in this is the concept of margin gains. With many substitutes from which to choose, even solely within the ex-farm category, issues can be raised on comparative effectiveness. There is also the question, one ingrained in agroecology, as to the percentage of an input that actually accomplishes the intended task and what happens to the missing percentage.

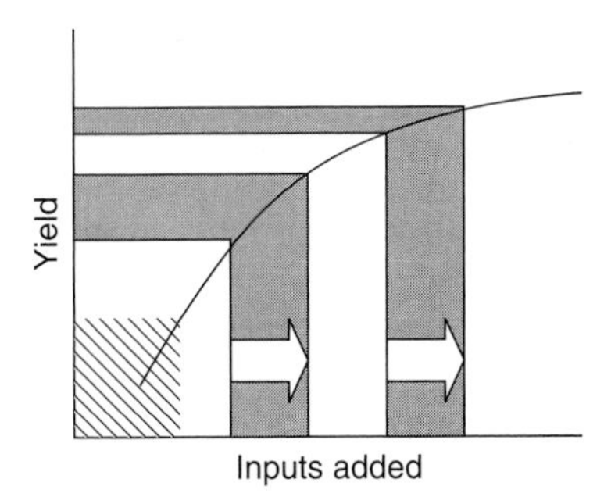

FIGURE 10.2. Marginal analysis where equal amounts of inputs (the two arrows) do not give the same result. This analysis determines when to stop adding inputs. The lower left, shaded box, box represents the pre-input, in-soil essential resources.

Marginal Gains

With marginal gains, small increases in the value of the inputs should result in larger increases in revenue, i.e., the costs of the input applied should not exceed the increase in anticipated revenue. This is demonstrated in Figure 10.2.

Figure 10.2 carries certain assumptions, all are true but, upon closer examination, not stalwart. One roadblock to complete picture lies in which resource hypothesis prevails (Figure 10.1). Statistically, this has proved almost impossible to determine.[6]

The problems of shifting resources dogs this breakdown. In multiple resource situations, the time of day, as well as the seasonal climatic variations, keep altering the balance, i.e., an abundance in one-time period may become a shortfall in another. For example, morning dew may eliminate moisture as the limiting resource for a short period, otherwise this remains. The net effect of two uncertainties, that of function and ever changing limiting essential resources, makes this a less than precise line of investigation.

Marginal gains, as in Figure 10.2, is the means to determine which mechanisms are substituted and, once selected, the amount employed. Whatever the options selected, it should offer more effectiveness and/ or cost less. This applies across the full ex-farm gamut. These are inputted until the per unit cost exceeds any monetary benefits gained. As expected, this analysis is tempered by environmental intangibles.

Truant Inputs

The relationship between inputs and outputs clearly exists, but often with considerable waste. For fertilizers, capture by crops is not 100%. The totals advanced are in the 40–60% range. For applied biomass, the capture number is about 20%.[7] Figure 10.3 illustrates this inefficiency.

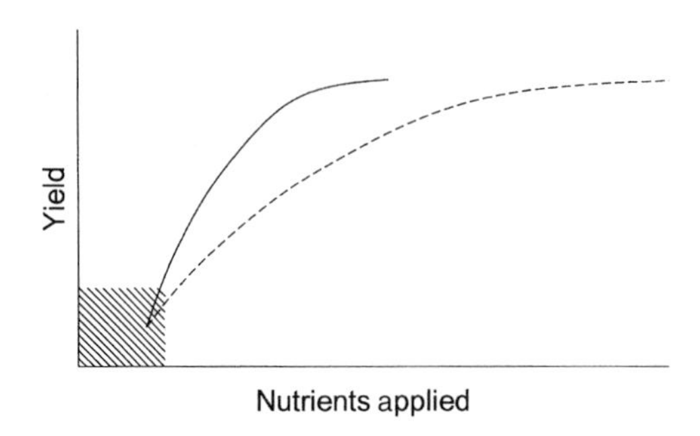

FIGURE 10.3. The classic resource in-output curve (solid line) as normally stated. The actual curve (dotted line) takes into account those truant resources lost to the system.

Lacking are the economics of loss, where those not captured, e.g., 40–60% of the fertilizer inputs, do not accomplish their assigned task. If residual to the system and available in subsequent years, there is little worry. If truant outside the system, these losses can lead to environmental deterioration.

Farmers want to direct, not over apply, expensive chemicals and this is where precision agriculture gains traction. From an economic and an environmental standpoint, this avoids a gross over application of fertilizers. Despite the uncertainties, some efficiencies are gained in keeping a correct balance between nutrient applied, those present, and light and water available. Spot applications, discussed later this chapter, are part of this.

Improving task efficiency is also a goal with herbicides, fungicides, and insecticides. The concept of integrated pest management (IPM) was developed to avoid saturating fields, with toxic chemicals at the first sign of a plant-eating insect. Instead, this approach monitors plots, applying insecticides strategically just before populations of damaging insect begin an upsurge.[8] Although an improvement on overuse, there is still the problem of losses outside the plot, through drift, and the resulting environmental danger.[9]

The amounts lost can be difficult to determine. If extensive, through a visual or other assessment, an agroecosystem re-design may be in order. What happens to wayward chemicals, and any environmental damage done, is a very understated aspect of farm economics.

DESIGN VARIABLES

As seen, applying the correct balance of fertilizer inputs is not an overly rigorous science. More can be done, not only through precision agriculture, but through agroecosystem design.

The mere presence of multiple species can insure a higher rate of in-system retention. This often occurs when each species has roots in different soil strata, the theory is that what one plant misses, the other will take.[10]

There are other strategies to capture wayward nutrients. Strips, specially formulated to take in whatever nutrients pass by help in this. Although of less value to current or even future crops, plot-external strips, often as riparian buffers, are the last preventative; before nutrient are totally lost and before these cause faraway environmental damage.[11]

Spot Applications

With multiple plant species comes the opportunity for greater application efficiency. Instead of blanketing an area, essential resources are placed in active root zones.

For intercrops, there can be efficiency gains if the nutrients are also so aimed. If two productive and co-inhabiting species are highly complementary, the application of nutrients is between rows of unlike species (Figure 10.4, top). If the two species are less complementary or even somewhat completive, then nutrients may be best reserved for each species by placing these between rows of like species (Figure 10.4, bottom).[12] Whichever course is taken, between rows of like or unlike species, this is a cost-oriented strategy, one intended to reduce fertilizer use by increasing conversion effectiveness.

A similar strategy exists with water. Drip irrigation, with or without in-solution nutrients, can be row supplied (as in Figure 10.4, bottom). This allows water to be applied according to each plants needs (i.e., volume and timing). The disadvantages lie in the costs, both in capital equipment expenditures and in daily operational outlays. Although miserly as to water use, these systems are reserved for situations where revenue-oriented productivity can be justified.[13]

Timing

With conventional timings, fertilizers are applied before planting or mid-season before plants begin to demand nutrient in the greatest amounts. There is another option, to restock the soils at periodic, rather than seasonal, intervals.[14]

Phosphorus, potassium, and trace nutrients can be taken faster than they are naturally replenished. Rather than directing nutrients seasonally to one crop type, the strategy is to add these when the in-soil supply falls below economically viable levels. This comes with an economic advantage. It can be cheaper to apply these during the off-season

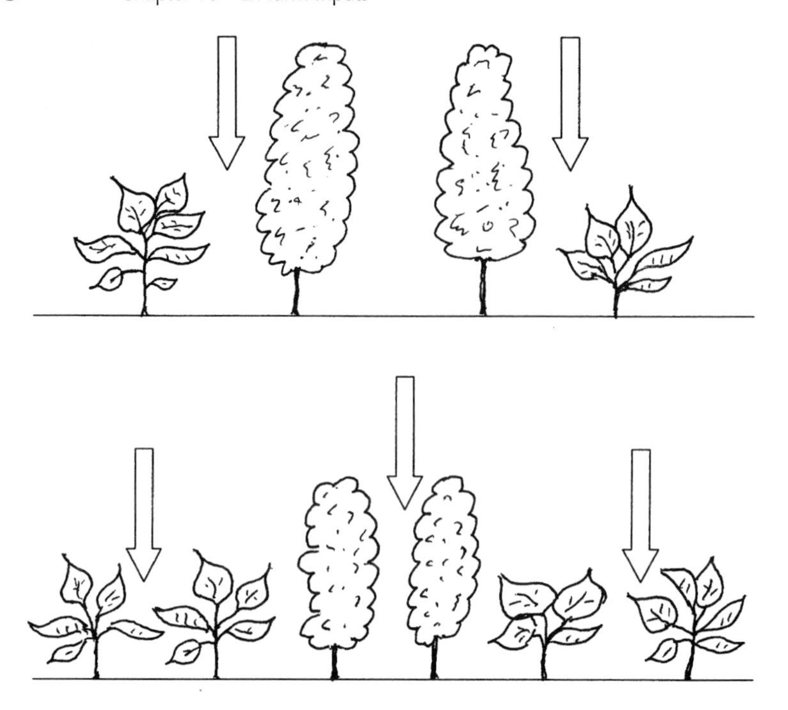

FIGURE 10.4. Two spot nutrient application strategies, the upper places nutrients between rows of unlike species, the lower between rows of like species.

when in-house labor and machinery are not in great demand. This strategy is only utilized when the nutrients applied stay on-site.

A replenishment strategy, pre- or mid-season, may incorporate suitable covercrop. The idea being that a cover species capture and hold the surplus elements until needed by the primary species. Release is through decay when the cover species goes into decline or dies. The economics of a fertilizer-covercrop strategy are far from certain.

Timing can be critical with insect outbreaks. It is better to apply minor amount of insecticides early rather than waiting and be forced into large doses. This is where IPM offers opportunity to catch a growing problem early. The ex-farm version monitors insect populations and locations, triggering a response when predetermined danger levels are detected.

ASSOCIATED AGROTECHNOLOGIES

The ex-farm vector is lacking in agrotechnologies, but rich in design variables. This is because much is one dimensional; few inputs offer enough ecological expansion and differentiating biodynamics to qualify as agrotechnologies.

Because of their physical presence, forms of irrigation such as spray and drip qualify. Also inclusive as an agrotechnology would be insect and rodent traps.

Irrigation

Irrigation, flood, spray, or drip, are clearly agrotechnology that are added-on to farm plots. These are mostly revenue-oriented.[15]

Flood irrigation, covering the ground with irrigation water can be wasteful. Cost of water aside, this is the cheapest of the irrigation techniques.

Sprinkler irrigation is less wasteful but a power intensive means to water crops. Spraying does offer a side benefit. The sprinklers, turned on during cold nights, do protect against mild frosts. The proviso being that the crops do not lodge if enshrouded with a coating of ice.

As previously mentioned, the most water efficient is drip irrigation. For this, a system of hoses continually drips water onto the root zone of each plant. This has proven successful in water conservation in that comparatively small amounts are required as this resource is well targeted.

Clearly, crop irrigation is an attractive option, both for insuring high yields and for reducing drought-related risk. Still, there are perils. The unwise use of scare water may drain natural surface water that supports plants and wildlife. Deep sources may not be limitless, the water may be a relic of a distant past or refill a slower rate than taken.

Other drawbacks are with soil–salt accumulation. More problematic with flood or spray methods, far less an issue with drip irrigation, slightly salty water can leave behind a residue. If allowed to build, this can poison the soil for crops. Another disadvantage of flood and spray applications occurs when excess water runs off; it can carry away nutrients for no good purpose.

Traps

Insect traps are a class of ex-farm add-ons that are very effective at countering herbivore insects. These can target certain species without harming the ecology of an area, irrespective of how crops are raised.

These can be preventative or employed solely in for insect outbreaks. Apple maggots, bag worms, corn earworms, European corn borers, fruit flies, and peach tree borers are part of a growing list of insects that can be trapped and removed through physical and/ or chemical-based traps. The most potent weapon, and lure, in this fight may be synthesized, insect-attracting pheromones.[16]

Less sophisticated lures and snares have long been a staple of the backyard grower. In the tropics, coconuts, each with a small hole, buried, can attract foliage-eating ants. Later the coconuts are taken away, the ants are killed and the coconuts repositioned.[17] Beer-filled sauces, placed at ground level, attract and drown leaf-eating slugs. Gardening books list further possibilities.

Traps also function against larger animals such as mice and rats. In some cases, this may be less effective as these creatures soon learn to avoid this danger. Still, traps with barriers used to thwart rice-consuming rodents has been found as cost effective as poisons.[18]

ENDNOTES

1. A study of fertilizers and soil health, one showing not so positive results, is by Sharma and Subehia (2003).
2. The counterproductive aspect of broad insecticides is in van der Valk *et al.* (1999), the issue of poor dosing, high or low, is in Kaiser (2003).
3. The resource hypotheses are discussed by Paris (1992) and Dai *et al.* (1993) with further development by Harmsen (2000).
4. To give a few in many such references; for wasps Starcher (1995), for ants Fuente de la and Marquis (1999), and for spiders Marc *et al.* (1999).
5. Milkweed against nematodes and armyworms is found in MacKenze (1999), milk against certain mold was researched by Bettiol (1999). A listing of benign chemicals is in Weinzierl (1998).
6. The problems of statistically deriving the von Liebig, the Mitscherlich, and other resource-use hypotheses are given in de Wit (1992), Paris (1992), and Dai *et al.* (1993).
7. The 40–60% fertilizer capture figure is from Matson *et al.* (1997), that of biomass is from Palm (1995).
8. A description of IPM and its original chemical related goals is found in Walter (2003, p. 59).
9. Inside plots, chemical fertilizers do, over time, harm the character of the soil (Sharma and Subehia, 2003). The same holds true with insecticides, herbicides and the like. In one case, residuals from past-used, metal-based fungicides (e.g., copper) interfered with soil algae (Zancan *et al.* (2006).
10. As endnoted in Chapter 4, rich sites have the potential, through marginal or conversion gains, to give the high land equivalent ratio (LERs). The resulting axiom, unproven but most likely true, is that, with moderate or high fertilizer applications, intercropping is the best option and should prove economically superior whenever crops are hand planted and hand harvested.
11. The different designs of riparian buffers and other nutrient capture strips are covered in Wojtkowski (2002, p. 233). The faraway damage alluded to includes dead zones in lakes and seas caused by nutrients from the farm runoff.
12. Fertilizers and the advantages of placement between rows of like plants, is discussed in Wijesinghe and Hutchings (1999).
13. Drip irrigation becomes less revenue oriented as the plants become larger (e.g., in an orchard setting) as this reduces the per unit (of output) irrigation cost.
14. The restocking of nutrients at period, rather than at seasonally directed intervals is an extrapolation of work done, and suggestions made, by Gosling and Shepard (2005).

15. Although clear-cut in most cases, the economics of multiple-user, free-flowing irrigation systems can become complicated, especially when farmers water both cost- and revenue-oriented crops (Guillet, 2006).

16. Pheromones in agriculture are discussed in Kirsch (1988), and Suckling and Karg (1998) and, in forestry, by Dedek *et al.* (1989).

17. The coconut-ant trap is from Brown and Marten (1986).

18. The economics of traps against rice-eating rodents is from Singleton *et al.* (2005).

11 Microbial and Environmental Setting

The argument can be made that the environmental setting and microbial vectors are highly interrelated. There is truth in this, the environmental setting often sets the stage for the microbial. This does apply in all cases, the vectors can and do operate with independence.

APPLICATIONS

One agroecological principle is to let nature do the work. Appropriate biodiversity and/ or a favorable landscape, coupled with the right environmental setting, can encourage insect-eating bugs to attack the plant-eating kind. This is just one manifestation. Insects do fall victim to diseases and these can be breed and spread for this express purpose.

There is a lot more. Although not explored in great depth, one can make the case for agriculturally favorable microbes, microflora, and microfauna and the favorable productive dynamics generated.[1]

The environmental setting insures that the needed organisms find farm fields a suitable home. This is sweeping topic that includes (in addition to facilitative microorganisms) earthworms, anti-nematodes, and insect-eating insects. This can be through single-task organism to something more elaborate, such as a mini-ecosystems with the larger agroecosystem. The latter has not one, but a series of contributing organisms linked through an environmental setting.

There is the much cited study of the citrus ant (*Oecophylla smaragdina*) in China. This is more that a single species, but is a mini-ecosystem or ecological matrix that materializes from a favorable setting. The site is an orchard monoculture of orange and pomelo trees. Bridges between trees, in the form of canopy-connecting bamboo strips, allow the ants greater freedom of movement. The pomelo trees help the ants over winter.[2]

An ecological matrix or hierarchy comes about as more that one insect type is involved, but also part of the citrus ant admix are mealy bugs, wasps, and ladybugs. The mealy bugs, although they are orchard destructive, provide nectar for the ants. As with many natural controls, some losses are tolerated for a greater good. In this case, it is the ants eating or driving off the more destructive pests. Although the ants are featured, the wasps and ladybugs target insects that the ants do not consume. The ants do not disturb the wasps and ladybugs.

This is illustrative of the types of interactions, far from superficial, that can be harnessed. These can occur at smaller dimensions.

The simplest version of the microbial vector is narrowly focused, microbes are only assigned a single task. These would be introduced, the success of their efforts is, in the least, dependent on a non-hostile site. Of course, matrix or hierarchies at the microbe level are more than possible and, through a strong environmental setting, these may exist unobserved.

A list of what can be done through either vector is quite long and further development is a distinct possibility. In accomplishing agricultural tasks, these are lines of development where one might truly expect the unexpected.

UNDERLYING BIODYNAMICS

Small organisms, microbes, micro-flora, and micro-fauna, either directly or indirectly fostered, are the basis for the microbial vector. The tasks expected include:

- adding those diseases that afflict plant-eating insects;
- more in-soil nutrients being captured or recovered, from the air or cleaved from rocks or entrapping chemical associations;
- better utilization of soil elements;
- moisture being infiltrated, held, and better utilized;
- plants better resisting temperate extremes;
- plants better resisting droughts.

Through the environmental setting, the tasks undertaken are:

- anti-insect insects with a better survival, breeding, and retention rate;
- a bloom in microbes that, taken together, accomplish microbial tasks (as above).

Yield Gains

A number of species have strong associations with microorganisms. Pines deserve mention in that they are aided by nitrogen-fixing, in-soil, mycorrihiza and, additionally, by other microbes that make available other, in-soil minerals. These associations may be more widespread than thought. In example, exotic eucalyptus, those grown outside their native areas, grow better when inoculated with microorganisms imported from their home ground.[3]

In increasing yields, one option is to seek soils, in particular, those microorganisms that have co-evolved with the species in question. This might be with pre-domesticated versions of common agricultural crops. Supporting this view is the understanding that many microorganisms are ecosystem specific (Cleveland *et al.*, 2003).

Also possible is the interspecies transfer of sugars, where one species indirectly increases the yields of another. This occurs at the fungi level.[4]

There are other contributors. Crops must be pollinated and insects offer this service. Honeybees are most notable, but other types and a few bird species prove their worth in this respect.

Cost Reductions

Given that microbes and other small organisms function on their own with few monetary outlays, the cost gains are unquestioned. Enhancing their effectiveness starts with land preparation and the best in this regard is no-till where no plowing is done. If not feasible, a tractor-pulled chisel tool will break the soil to some depth, allowing roots to ease of penetration but with less of an impact than the more intrusive methods.

Unless the environmental strategy of no-till or a pulled chisel tool includes anti-weed measures, these methods may not be cost savers. Moldboard or disk plowing have weed control possibilities (see delayed sowing, Chapter 6) and do maintain a favorable environmental setting. For one, these do not grind up the earthworm populations as occurs with rotary plows. Moldboard and disk plows are also cheaper to operate than rotary types.

As for insect-consuming insects, habitat strips are better mowed with less intrusive and less costly machinery. It does not help if, in the grass-trimming process, the desired insects are pulverized. The best course is to cut the grass and herbaceous forage at the stem base. Tractor-mounted side cutters do this. Again, the simpler the process, the less costly it should be.

Risk Abatement

Risk may be large favor in both the environmental setting and the use of microbes. The potential is so great that these could, given the right choice of the other vectors and agrotechnologies selected, supplant ex-farm inputs. A large portion of this comes in reducing risk.

Water Dynamics

Microbes can improve the ability of the soil to hold moisture. Earthworms do more, enriching the soil-eating and excreting organic materials and, through channels cut in the upper strata, improving water absorption. These creatures, in providing in-soil tunnels, also hasten the spread of microorganisms.[5] Even insects thought of as crop detrimental, such as the termite, improve soil structure and the ability to infiltrate and hold water.[6]

When water is in short supply, fungi can be enlisted to better resist this shortfall. In one study, water-stressed wheat survived 10 days longer when inoculated with a facilitative fungi. As for the environmental setting, manure is longer lasting and holds water better than fertilizers. The till method and use of rotations also helps in this regard.[7]

Insects and Plant Diseases

There is a large and growing body of what may be loosely termed bio-insecticides, inclusive under this heading is the introduction of insect-fatal diseases. The most prominent of the leaf-applied commercial insect pathogens is *Bacillus thuringiensis* (bt). Another purchasable is milky spore disease (*B. popilliae* or *B. lentimorbus*). This stays in the soil, infecting generations of in-soil grubs. Other facilitative microbes include *Pasteuria penetrans* which prevents nematodes from laying eggs and *Trichoderma hazianum* which shields plant roots from invading fungi.[8]

For farms with few resource, there is a less sophisticated form of biowarfare. Dead or dying caterpillars are squashed and the juices, with water added, made into a diluted solution. The notion being that, through on-plot spraying, the disease spreads, killing the targeted bug, harming little else.[9]

Plant diseases can in turn fall victim to disease. The blight that attacks the European chestnut and has almost destroyed the American chestnut can succumb to a blight of it's own. These plant disease-afflicting diseases are another avenue of approach, one less tried.

Temperatures

Endophytes, fungi living on or in plants, can help crops withstand extremes in temperature. As examples, watermelons and tomatoes have been shown to tolerate temperatures 10°C higher when furnished with an appropriate fungi.[10] Again, this is an approach with considerable potential.

Relevant Guidelines

How microbes and spores are best employed is an open question. These can be treated as an ex-farm, applied as needed. Better yet is to have these present and continually active in the soil or aboveground surroundings.

It may be best if the setting triggers a cascade of organisms and positive agroecological events. This can be the keystone species, as in ecology, where one species set the stage for many and, ultimately, for a flourishing mini-ecosystem.

For machinery use and other land management methods, the general rule is that, the lighter the touch, the better the result. There may be exceptions, but in seeking a least cost solution, the less is better approach should be the first explored.

ECONOMIC MEASURES

The effectiveness of single or hierarchal organisms, either macro or micro, are through established indices, mainly variations of the land equivalent ratio (LER) and the cost equivalent ratio (CER). Restating the latter, in somewhat different form,

$$CER_{(LER)} = \left(\frac{C_{ab}}{C_{ab1}}\right)\left(\frac{Y_{ab1}}{Y_{ab}} + \frac{Y_{ba1}}{Y_{ba}}\right) \tag{11.1}$$

This compares two designs or treatments (ab and ab1) for an intercrop of species a and b.

For resolving differences in treatments and design, straying from the norm by employing, as the denominator, non-monocultural values (as above) is possible. Because the standard of comparison here is an intercrop, rather than a monoculture, equation (11.1) does not produce a stand-alone value. This only occurs with values from equations (5.2), (5.3), and (10.1).

DESIGN VARIABLES

Microbes, in their most basic application, are design variables. In addition to the identified organism and a targeted problem, the approach must be field applicable; including the timing, method of application, and in which form these are best applied. With few exceptions, this remains a topic in need of further development.

ASSOCIATED AGROTECHNOLOGIES

Much of that presented in this chapter hugs a fuzzy line between the design variable and the lowest configuration of agrotechnology, that of the add-on. There are a number of agrotechnologies, undoubtedly others will be put forth in the future.

Composting

The notion behind composting is to provide essential resources by biodegrading plant materials off-site and employing these as a garden or farm input. Through this input, agroecosystems are not only provided with mineral nutrients and water-holding organic materials, but well seeded with beneficial, and maybe a few less beneficial, organisms. These include earthworms and multiple species of microbes. As this requires a large amount of labor, composting is generally regarded as revenue oriented.

Tillage

How the earth is prepared for planting sets or upsets the stage for host of factors. As a one-dimensional process, simply selecting the right plowing technique for the crop, e.g., deep plowing for deep-rooted cotton, this is a design variable, not an agrotechnology.

As discussed, the choices under tillage are a turning of the soil (either through moldboard or disk plow), rotary tilling (through a rototiller type device), chisel plowing (with a deep, narrow blade), or doing nothing (no-till) prior to planting. There are also hand methods and associated variations such as ridge tilling where mounds are shaped and later planted upon.

The economics is much dependent on the equipment farmers have available. New machinery may be too expensive for what may be minor ecological gains. If machinery needs to be purchased, an environmental setting should be inclusive in the decision.

Beehives

In the larger scheme, one must not forget pollinating insects. Their effect on yields are well recognized. Beehives and bees are independent from the mainstream agrotechnolgies and might be considered an economic add-on and a source of income. As a side benefit, well-placed hives can be positioned to drive off large grazing animals.[11]

Bird and Bat Houses

Another approach with great potential are birds and bats. As major consumers of insects, these provide a very promising avenue of control. This is a case where the dynamics of the natural ecosystem are harnessed for agricultural good. As a means to reduce inputs, birds and bat largely fall under a cost orientation heading.

It is nice when a passing bird ingests a few herbivore insects, it is better yet if they stay around, consuming harmful bugs in large quantity. By offering birds a suitable home, they may take up residence in the immediate area.

Agricultural bat or birdhouses of proper design and location are part of the plan. Since insect infestation can be passing event, feeding stations, i.e., bird feeders, are provided when insects are in short supply. Feeding is suspended when a pest management survey indicates that the damaging insects are exceeding the desirable base populations. The feeding stations are also used to attract the birds to concentrations of harmful insects.

There are as many options along this line as there are insect-eating bird species. Wrens eat large quantities of beetles, especially when they have nests of young to feed. Well-designed and positioned birdhouses, coupled with a favorable habitat, can keep wrens on-site.

More can be done through mini-ecosystem or control matrix. Some birds pick insects off leaves, others grab them in flight. This is an example of niche feeding that can be exploited for increased efficiency. Domestic fowl, chickens, ducks, turkeys, etc., pick insects off plants, those trying to escape may be captured in fight by swallows or swifts. Bats seek out and ingest those bugs that take wing after dark.

There is more potential. Mice and rats, at times a cropping problem, are eaten in the daytime by hawks, eagles, and, at night, by owls. If provided with birdhouses and roosts, these can be more effective control strategy than easily bypassed traps.

Parkland systems, with their scattered trees, provide the hunting roosts and/ or resting places needed by rodent-hunting owls and insect-eating bird species. Bat and birdhouses are an add-on to the parkland agrotechnology that helps to keep agriculturally favorable birds and bats on-site.

ENDNOTES

1. The view that microbes are underrepresented in the literate comes from Klironomos (2002), that these can dominate the agroecological scene from Reynolds *et al.* (2003).

2. The citrus ant and the promotional habitat surroundings are in Huang and Yang (1987).

3. Part of the discussion on microbes and soil enrichment is in Setälä (2000) and inoculation by Garbaye *et al.* (1988). A general discussion of N-fixing mycorrhizae is by Picone (2003).

4. The interspecies transfer of sugars by fungus is from Pennisi (2004).

5. Microbes to hold moisture are discussed by Statälä and Huhta (1991) and the benefits of earthworms by Hulugalle and Ezumah (1991), Hauser *et al.* (1997), Scheu and Parkinson (1994), and Shuster and Edwards (2003).

6. Ants are reviewed by Stanton and Young (1999) and Molles (1999, p. 338), termites by Mando and Van Rheenen (1998).

7. Pennisi (2004) mentions water and microbes, the effects of manure are from Kihanda *et al.* (2004), those of till and rotations from Gregory *et al.* (2005).

8. Milius (2003) discusses relationships between microbes and disease, the two examples are from Millman (2005).

9. Biowarfare against harmful insects is described in Smits (1997).

10. This is from Pennisi (2004).

11. Honeybees against large animals, in this case elephants, are mentioned in Anon. (2002) and O'Brien (2002).

12 Single-Plot Design

Economic analysis does not stop at, but begins with the individual plots. This might be accomplished through raw vectors, subdividing into agrobiodiversity, biodiversity, rotations, etc.

Plainly, not all need, nor want, an abstract approach. Farmers and extension agents often desire something more practical. This can be in the form of a strategy, means, and/ or method prepackaged for immediate and effective use. The agrotechnologies do this. These are less abstract, far more practical, providing an immediate avenue for implementation.

Throughout the previous chapters, the more common agrotechnologies are presented.[1] Use can be as simple as the selection of a principal-mode agrotechnology. More often, a mix of agrotechnologies are incorporated. Terraces can support seasonal intercropping and a parkland can be, and frequently is, added to the mix. The problem is more of compounding than being a single choice.

TASK ORDERING OR LAYERING

When dealing with nutrients, weeds, erosion, water infiltration, plant-eating insects, plant diseases, destructive animals, etc., some vector mechanisms confer a strong effect, others are weak; some are long lasting, some transitory; some function best if reenforced, others are a solo effort. Hand weeding is an example of a strong and transitory anti-threat mechanism. Not self-sustaining, it must be done over and over.

If one strong mechanism is sufficient, and economically viable, then nothing more need be considered. Of course, threats come in different forms and in varying magnitudes. This is demonstrated in the wide range of plant-eating insects and plant-afflicting diseases. All or most of which must be combated.

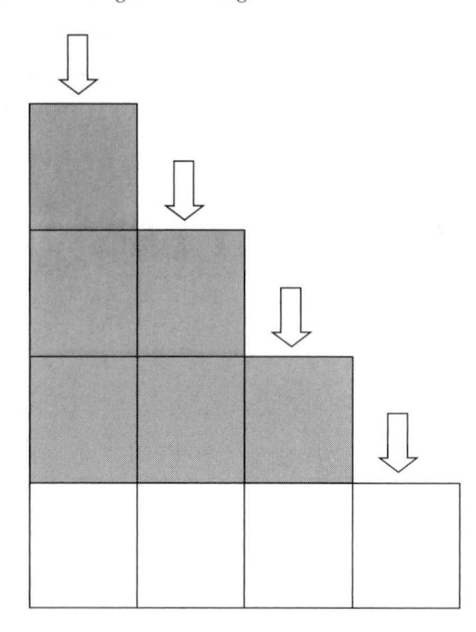

FIGURE 12.1. The different agroecological options. Top to bottom, these boxes represent a series of controls (left side) or a single counter (right side). The shaded boxes are permanently in place, the unshaded boxes represent mechanisms brought in to counter an emerging threat.

In eschewing harsh chemicals, the next course, especially when dealing with groups of threats, is to employ multiple mechanisms. Not implemented willy-nilly, a plan is needed. A visual representation, Figure 12.1, shows how the threat countermeasures are layered.

Each block in Figure 12.1 is a mechanism or means to satisfy a need or counter a threat. On the left side, the stacked, vertical blocks represent a series of countermeasures. The countermeasures are engaged such that each is mutually harmonious with another, each providing a separate layer of protection. The single block on the right side is for a single mechanism. This can be a broad-based insecticide that kills any emerging insect threat or a hand weeding to remove all unwanted plant intrusions.

Taken as a top-to-bottom series, these ecological tracks represent an integrated, complementary approach to insect, weed management (e.g., tillage, rotation, and certain crops) or multiple mechanisms or treatments to keep soils fertile (e.g., biodiversity, rotations, and added biomass). Usually the upper blocks (shaded in Figure 12.1) are permanent, always in place, always functioning, e.g., anti-erosion ditches. The lowest (unshaded blocks, Figure 12.1) are transitory, treatments applied as and when needed.

This relative placement of the tracks depends on those mechanisms or treatments available. To take life, a separate figure is required for each need or threat and each block is so labeled. Once all is in place, the agroecosystem design process selects which vertical track is the best option via the full range of system threats and requirements.

Essential Nutrients

Soils require essential elements, trace or otherwise, if they are to yield. How these are acquired is an integral part of agroecology. Within the framework of Figure 12.1, there is the right-side, single application or the left-side, multiple methods. Assigning mechanisms, three vertical tracks, i.e., left, right, and center, might be formulated, as below:

Rotations with fallow		
Agrobiodiversity	Rotations without fallow	
Off-season facilitation (e.g., covercrop)	Cut-and-carry biomass	Fertilizers

The above could represent the best means to increase the amount of a limiting elemental resource. There would be similar tracks for other mineral resources. In actual field use, the blocks will be better defined, including which plants are to be paired and the exact makeup of a rotational sequence.

In the above scenario, the three option sets (tracks) are mostly revenue oriented. Knowing the economic orientations within each track, along with assigning a land equivalent ratio (LER), revenue, and cost values, are part of the process.

Water

If rains are infrequent, tracks can prove useful to insure that crops have sufficient water. These are:

Windbreaks		
Contour ditches	Low transpiration crops	
Drought resistant crops	In-soil biomass	Irrigation

As an added note, the above, left and center, are anti-drought formulations. A similar vertical tracks might be formulated as anti-inundation or anti-flood countermeasures.

Insects

In dealing with crop-eating insects, this can be done in one swoop (as with the application of a broad spectrum insecticide) or in multiple steps (with layered or stacked natural controls). This can involve insect maintenance, i.e., keeping populations below the level where they are doing any significant damage, or, if maintenance fails, simply attacking the growing threat.

Figure 12.1 again applies, this time to the insect control options. The upper, permanent, a general or non-specific control, the shaded blocks, are maintenance. The lower blocks can be broadly targeted, benign or harsh chemicals and be aimed at whatever insect pest is threatening at a given moment. Where available for purchase, wasps and ladybugs can also be a lower block countermeasures, all allied against a specific threat. Traps are another such option. A series of anti-insect vertical tracks may be:

Between or in-plot strips
In-crop repellant plants   Bird habitats with houses
Introduced predator insects   Spot-applied remedies   Insecticides

The number of insect control options is much greater than with other threats. This introduces more flexibility and opportunities for more than three tracks of solutions. It is possible to stack four or five layers of protection, again permanent to transitory, generalist to outbreak.

Weeds

With weed control, the strategies are less understood with fewer in-place options. There is single, right-side application, either hand weeding or a broad herbicide. The multiple methods are the type of till, crop rotations, intercropping, cut-and-carry mulch, and the treatment of crop residues (e.g., fire). As with insects, there are maintenance countermeasures, but less in the way of outbreak control:

Rotations
Tillage method   Suppressant mulch
Covercrop   Hand weeding   Herbicides

Temperatures

The control of plot temperate tends to be more in-place, but with some immediate countermeasures. Of the possible tracks are:

Well-placed stones
Protecting biomass   Parkland trees
Microbes   Resistant species   Water spray

Erosion

There are other threats and anti-threat countermeasures. The soil erosion countermeasures can range from permanent to transitory. Although the chemical options, such as a soil-binding agent or soil sealant, are less recognized, the possibilities are still there[2]:

Ditches or bunds		
Dense biodiversity	Agrobiodiversity	
Covercrop	Cut-and-carry biomass	Soil-binding agent

MULTI-TASKING

Taking the easier course, that of a single, right-side counter, is an attractive option, one too often taken. The alternatives, the multi-track solutions, are more difficult to implement, especially since a different tracks address different needs and threats. The multi-track key lies in the detail, identifying specific shortcomings or looming dangers, and plugging the gap. The dangers can range from one insect pest to some weather event.

Applied Tasking

To illustrate the complexities of multi-tasking, it may be best to lead through example. Take hillside agriculture where periods of rainfall are brief, but intense. In extracting the most moisture, there is more than one solution. Two anti-erosion mechanisms are:

(a) larger plots of facilitative biodiversity (with an understory covercrop), bisecting the plots are slope-crossing infiltration ditches, or
(b) narrow, monocultural strips that contour the slopes.

The first (a) represents a two-track solution to the water problem. The second (b) is single stage. Interesting, these can be combined into a vertical track of three blocks (as in Figure 12.1). This reenforces what could be, depending on the site, 2-weak options (a and b above).

More can be done, both solutions have an anti-insect aspects. The first (a) offers refuge for herbivore insect in the covercrop and along plant-filled ditches. This again is a two-block track solution. The second (b) is a bit weaker with regard to insect spread deterrence, but not much more. Together, the effects are cumulative. Appropriate add-ons, say bat or birdhouses, would strengthen this further. As a final, lower tier defense, there is always the option of some outbreak measure, e.g., introduced predator insects.

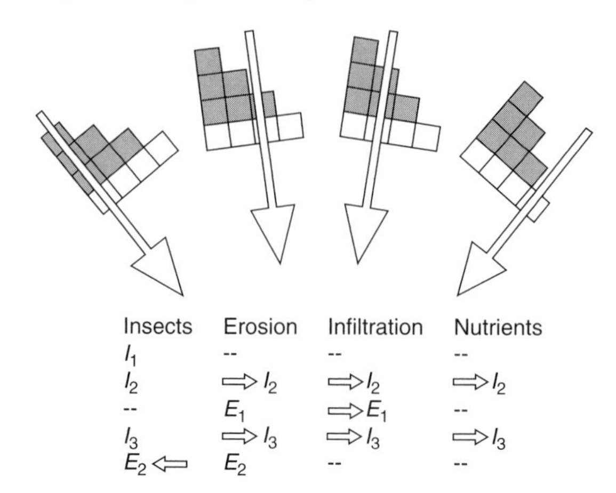

FIGURE 12.2. Multiple triangles in the design process brought together to determine which of the individual tracks are best employed.

As for weeds, the first (a) has better desirable plant characteristics (DPCs) due to the covercrop. Other add-ons, such as till and rotations, would reenforce the anti-weed potential.

Both solutions are suitable for farm machinery. Both can be further augmented on a number of fronts, one option being windbreaks to help conserve water and as a refuge for insect-eating birds (combined with birdhouses). It is this mix of the two approaches, with add-ons such as windbreaks, as illustrated in Figure 12.3.

Analytically, this can be expressed extracting the operative horizontal and vertical tracks from counter defenses (as in Figure 12.2). For the above case, this would read as:

Anti-insect	Anti-erosion
I_1 – bird refuges (with birdhouses)	E_1 – ditches
I_2 – vegetative strips (insect-eating insects)	E_2 – crop strips
I_3 – covercrop (insect-eating insects)	

More counters can be added, water infiltration (W) and nutrient replacement (N) are common possibilities. In matrix form, this is:

Insects	Erosion	Infiltration	Nutrient
I_1			
I_2	$\Longrightarrow I_2$	$\Longrightarrow I_2$	$\Longrightarrow I_2$
	E_1	$\Longrightarrow E_1$	
I_3	$\Longrightarrow I_3$	$\Longrightarrow I_3$	$\Longrightarrow I_3$
$E_2 \Longleftarrow$	E_2		

FIGURE 12.3. A cultivated hillside in cross-section showing a composite of mutually reenforcing agrotechnologies. Represented are windbreaks, contour crop strips and anti-erosion ditches.

The idea here is to check the potential carryover, i.e., how much can be accomplished with one influence. Above, I_2 (vegetative strips) also helps horizontally with erosion control, water infiltration and, through root pruning, providing nutrients to crops.

Each column can be summed vertically to check the effectiveness in the assigned tasks, the total cost of the intervention or its effect on revenue. Not to be overlooked, effectiveness can also be gauged using risk analysis (the risk index).

In deciding how many of these ecological mechanisms to add, the easiest economic measurement is the cost–benefit of substitution, i.e., what could be done compared with what is.

In the matrix, I_2 offers gains in insect control as well as erosion, infiltration and nutrients. This sum might prove more valuable than single purpose I_1 as the cost is amortized across multiple substitutions. The ultimate decision on when tracks are best will depend on (a) how many countermeasures are best, (b) how much yield potential is lost (through area taken or a now higher base level of herbivore insects), (c) the individual task effectiveness of a countermeasure, and (d) their combined effectiveness.

Marginal Effectiveness

Complicating the analysis, the effectiveness of each mechanism is subject to marginal gains or, in this case, marginal effectiveness. Figure 12.4 (left to right) shows how the performance of each declines as the application intensity increases. For example, more insect-seeking birds means that each must spend more time searching for meal, proportionally less time eating, a decline in per bird marginal effectiveness.

With in-plot strips, this often means that, as the strips broaden and cropping area decreases, these reach a point of maximum effectiveness beyond when marginal gains decline. This implies an optimal.

For strips, there must be a balance in revenue lost through area taken against the gains in not having to employ expensive chemical controls. For birds, their low cost (say in birdhouses erected) means that, in economic terms, the number is only limited by natural dynamics, a farmer does not want them going elsewhere in search of an insect meal.

A second factor in marginal effectiveness is the additive effect as additional mechanisms come into play. There will be a potency ordering, some mechanisms dominating others. Birds, as a control, will trump predator insects simply because birds will consume a percentage of predator insects along with the herbivore types. In another example, certain predator insects may find a covercrop more to their liking making strips less of a factor as an insect countermeasure.

When the mechanism is generalist (targeting all insects), an ordering will evolve as to which mechanism dominates, which subside. When each mechanism targets a specific type of insect pest, there is greater latitude for intermechanism coexistence.

Without study, there can be a tendency to abandon one or more because, in unison with others, they appear ineffective. It must be remembered that control mechanisms do fail. The fallacy is in not looking into their importance as a backup or stopgap mechanism. For example, birds (I_1) may suffer a bad year and the insect control burden cedes to mechanisms I_2 and I_3 and even E_2. The lesson is to keep multiple in-place mechanisms as these can be prone to the whims of nature, increasing or decreasing in their cross-seasonal effectiveness.

The wavy curves in Figure 12.4 demonstrate yet another analytical complexity. Being exposed to climatic vulgarities, the interplay of natural forces, and inconsistent natural crosscurrents, one cannot expect smooth curves. Other reasons for lumpy or wavy curves are advanced later this chapter.[3]

In Use

A second case study, that in Figure 12.5, expresses much the same in terms of multiple tasking. In overview, the wind reductions of a windbreak/ border are augmented by standing parkland species. Also, insect-eating birds, nesting in the windbreak, can better ply their trade if they have scattered trees from which to seek prey. This is a

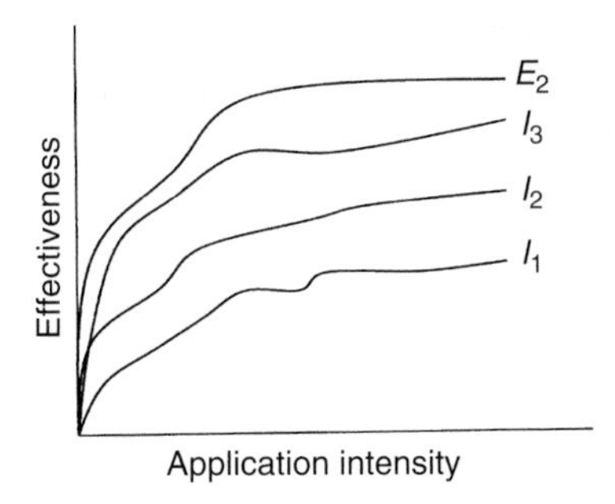

FIGURE 12.4. Declining margin gains as one or more different ecological control mech anisms ($I_1 - I_3$ plus E_2) are appended. This brings questions as to optimal application, ordering, and how many should be utilized. The wavy curves are explained in this chapter.

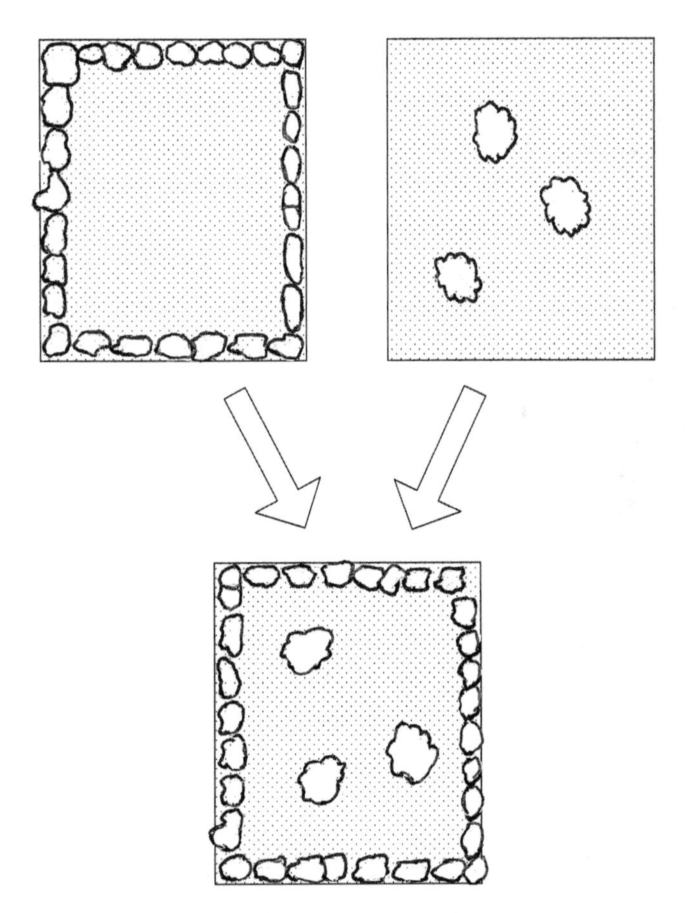

FIGURE 12.5. The merger of two agrotechnology. Here a parkland and border are combined for the sum of their individual agroecological effects.

practical expression of what happens in Figure 12.2. Again, the goal is to strengthen weak points, the desirable agroecosystem properties (DAPs), in the proposed agroecosystem, doing this through mechanisms that overlap in the ecological tasks they perform.

For the economics of Figure 12.1, those mechanisms as represented by the shaded boxes are inclusive in the pre-implementation economic analysis, i.e., the yearly cost of these permanent controls, when added to those of the crops, must project a profit. If one or more of the mechanisms are too expensive, another formulation is tried. In Figure 12.5, the boundaries and parkland trees, when added to moneymaking cropping system, must increase, not decrease, profitability.

As protections come from the in-place mechanisms, the lower, unshaded boxes (Figure 12.1) are a stopgap measure, actuated when the other track mechanisms are overwhelmed. As these are not expected to be used, these are not part of the pre-implementation analysis. The reasoning is that, when a threat is imminent and the crop is growing, the decision is whether it is profitable to save all or a portion.

For this, one must keep in mind that, at any stage in the growing season, the previous inputs are sunk, unrecoverable costs. The pre-period outlook was for a profitable growing season. Nature may have other ideas.

Midway in the season, the analysis, apply a transitory treatment or not, starts anew. The analysis would only look at those costs that occur after a problem occurs. If insects have consumed 50% of the crop, the question is 'is it profitable to save the remaining half?'. In making this decision, it must be remembered that the per unit costs of a treatment will have doubled. If deemed profitable, a transitory treatment is applied, if not, a total loss is accepted.

At each step along the cropping timeline, the decision process commences anew. If the determination was made not to apply the treatment, the farmer, in viewing that which remains at reaping time, decides if there is enough for a profitable harvest. That already spent is not inclusive in this decision. If the after-harvest income is projected to be below the cost of the harvest, the residual crop is abandoned.

DESIGN ECONOMICS

In designing agroecosystems or changing existing systems,[4] landusers do so with an eye toward increasing revenue while reducing both costs and risks. Complications come as, in aggregate, this is seldom possible.

The first step is to insure that the change is input efficient and in line with farm needs. Ratios, vector guidelines, and economic orientation, along with practical factors, are all part of the determination.

Financial analysis, using actual monetary values, also plays a role. No matter how favorable the LER and versions of the cost equivalent ratio (CER), a system should turn a profit. The agroecological and financial measures should agree, however, it is best to check.

Principal-Mode Agrotechnologies

All agrotechnologies are underwritten by one or more principal-mode agrotechnologies. A case in point, a two-species seasonal intercrop may be ecologically supported by rotational series and/ or by a parkland system. With this, the user expects to reap the ecological advantages of the three agrotechnologies.

Taking this case further into analysis, one expects that the intercrop, being at the core, is resource efficient. The first step in an economic breakdown is to look at the ratios.

Ratios

One of the essential elements of agroecological economics is ratio analysis. The LER is a measure of land-use efficiency and, by being intuitive, a means of cross-agroecosystem comparison. The value-added version (the relative value total, RVT) gauges input efficiency but with market prices included.

Continuing along comparative lines, the CER is a measure of input-use efficiency. The risk index (RI) looks at the yield stability of agrosystems and their proclivity for all or partial failure.

The strength of these are their intuitiveness. One does not have to ask the context, the values presented stand alone. Either the LER or CER must offer clear gains. The RI is a less explored, but still consequential, form of analysis.

There is an overriding reason for starting the process with ratios. The goal is not a sub-optimal agroecosystem, but one that is as resource efficient as possible. This can be with essential nutrients, labor, or any other on-site or imported input. Since on-site resources are cheaper to use, this is the initial focus.

Guidelines

Another check is that proposed systems operate within suggested guidelines. Exceptions are possible but, all-in-all, the best LERs, CERs, or RIs come about under the conditions suggested by the use of guidelines. To restate one key guideline, agrobiodiversity is usually best on weather-favorable sites with well-endowed soils.

The list of these is long and detailed. The importance here lies in insuring that the right vector combinations are being utilized.

Practical Considerations

Farms operate within certain parameters. Those with agricultural machinery should choose machinery-friendly agrotechnologies, e.g., strip cropping. Some agrotechnologies are best suited for farms that depend on hand labor.

Along these same lines the type of machinery is important. Treerow alley cropping needs narrower, more nimble tractors and harvesters and, if the current machines are inappropriate, a treecrop alley system, one with model spacing, is not a good choice.

Add-Ons

In multiplying agrotechnologies, the situation begins to resemble the left side of Figure 12.1 far more than the right side. If this is occurring, the problem is to insure some degree of efficiency and harmony. Agroecology offers add-ons. These can be large, as with rotations, or small, in the form of birdhouses and insect traps.

Cross-Harmony

Cross-harmony within the options chosen can be critical. One does not want birds eating all the predator insects, leaving only the plant-eating types. This is one reason for scrutinizing the underlying ecology. The best case is to overlap the tasks performed by each mechanism to obtain a high degree of reenforcement and each agroecosystem is well protected against the common and not-so-common adversities.

Economic Orientation

In introducing Figure 12.1, the tracks are mostly revenue oriented. If a hillside plot is located far from a household, chances are that a cost-oriented design is preferable. If cultivation is on scarce, well-watered, fertile bottomland, this will require, in all likelihood, a highly revenue-oriented agrosystem.

In putting mechanisms together into tracts, the economic orientation possibilities can figure into this. As shown, some plots are better revenue oriented, some better cost oriented.

Single, right-hand events tend to be a revenue-oriented addition. When compiling an agroecosystem, one tries to keep in a uniform economic direction, i.e., not mixing revenue- and cost-oriented additions. This is because there is no assurance, due to unforeseen cross-effects, that mixing result will always give the most profitable outcome. There is less chance of this happening when ecological processes go in one economic direction.

Input Efficiency

When dealing with a uniform input, such as an agrochemical (Figure 12.1, right side), the response curves are generally smooth (also as in Figure 10.1). This is not the case when mixed ecological mechanisms are involved.

When entering the realm of the multiple mechanisms, hops along the Mitscherlich-Baule surface (Figure 12.5) can easily produce a lumpy curve. However, this also occurs when there is a mix of vectors, each of varying strength, each producing uneven effects across their respective ranges where each interacts with another in different ways produces an uneven outcome (Figure 12.6).

The wavy curve is evaluated much the same way as the smooth version (Figure 10.1), except the solutions are less exacting. The starting point is the upper mechanisms in the triangle. In theory, those selected are the cheapest to install and manage and offer most in cross-effects. Instead of substituting for one input, say an insecticide, a mechanism might both replace an insecticide and reduce fertilizer needs. As shown (Figure 12.2), multi-tasking increases the overall value of mechanism.

As theory suggests, counter mechanisms are added until, at the highest threat intensities, the diminishing marginal effectiveness no longer support ecological additions. The mechanisms chosen must be

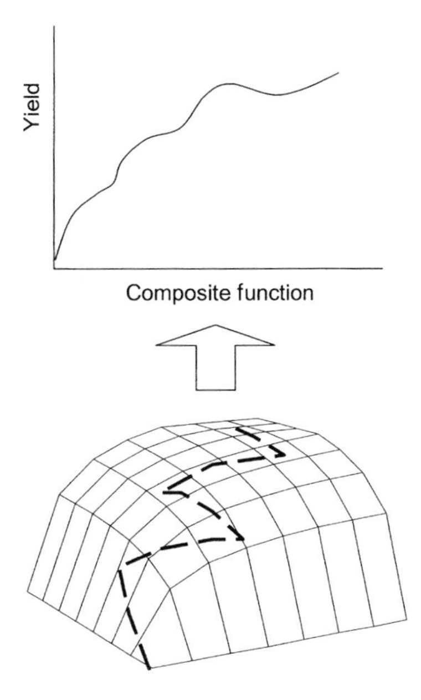

FIGURE 12.6. The lumpy or wavy input–output curve as derived from bouncing along and upward on the multi-resources Mitscherlich surface (as derived in Figure 10.1).

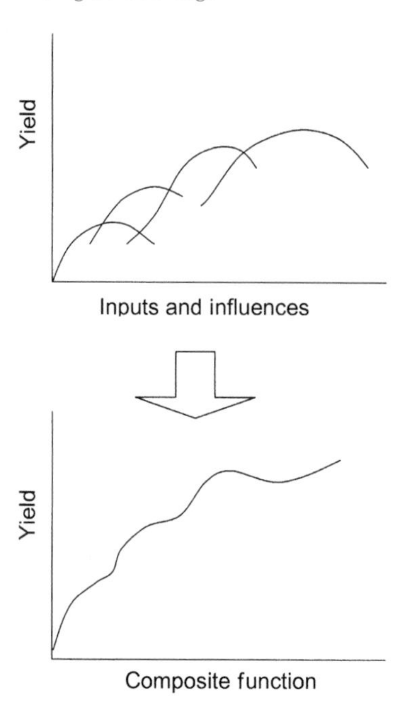

FIGURE 12.7. A second explanation for a wavy input-output curve. In this case, multiple and independent mechanisms, as from the options triangles, come together to produce this result (as exhibited in Figure 12.4).

effective across the full range of threats. Those permanently in place must function well in handling day-to-day variability, the transitory mechanism, those brought in when threat level rise, must be affective in countering a sporadic rise in the threat level.

A preference is given to cheaper, more effective mechanisms, those driven through natural biodynamics, those less dependent on ex-farm inputs. In doing so, there should not be a predilection toward simplicity but a tendency toward multiple countermeasures. The breakdown into permanent and transitory mechanisms should help in this regard.

Post-design Screening

As is often the case, the proof is in the application. Putting a plan into practice, hidden interactions and complications do arise that may waylay the best-designed agroecosystems. This is often a degree of trial and error that governs the design course of any active agroecosystem.

Profit

In examining a potential agroecosystem, profitability may be first on any want list but should not be the final word. Take the case of a highly profitable, but erosion-prone system. Despite the profit potential, it would be inadvisable to recommend this for placement on a loose-soil hillside. The option always exists for a redesign, keeping the same crop, but adding erosion-controlling features.

Environment

Not to be forgotten, by choosing threat counter mechanisms that are environmental helpful, an agroecosystem design can be so fortified. There are many examples. In combating soil erosion, water runoff is cleansed. In combating herbivore insects, the stage can be set for thriving communities of plants, birds, bats, and the like.

Culture

Another aspect of agroecology is cultural.[5] Often ignored, there are differences in how societies view the nature-agricultural interface. This is a broad topic, much of which is hidden from view. The importance is noting a cultural prohibitions or cultural preferences that can hinder or promote agroecosystem adoption.

Adoption

As seen, profitability is a strong lure, but there are other considerations. This comes together in one format. The in-use success of an agrotechnology may depend on matching the DAPs with the site and user objectives. There are many documented cases showing DAPs to be critical in selecting adoptable agroecosystems.

Hedgerow alley cropping finds greater acceptance where (a) maize is the primary crop, (b) soils are poor or declining, (c) rainfall can support this system, (d) productive intensity in increasing, (e) there is labor availability, and (f) erosion is a hazard. For live fencing, the preconditions for adoption include (a) animals running free near (b) high value market gardens, and (c) plots with no fencing or there is a need to replace dead fencing.[6]

CASE STUDIES

At first glance, this design process can seem overwhelming. This is not always the case. As expected, some situations are fairly simple, others far more complex.

As with any such process, the first chore is an accurate diagnosis. The simpler situations often involve a one-for-one substitution, switching one design variable (a spatial pattern, a species, the timing, etc.) for another. The result can be more than adequate if the problem is narrow in scope. Others, those less easily resolved, require looking at alternatives, including a shift to another agrotechnology.

Diagnosis

European farming in the medieval period provides an interesting diagnostic exercise. During this pre-agrochemical period, much agroecology was in place but, because some key vectors were not researched and underdeveloped, the overall system was less than successful at least when measured against current yield standards.

The rotation was a mainstay of the period. Soil health and outputs were maintained partly through strict, pre-planed cropping sequence, partly through the inclusion of a fallow. Looking at the overall practice in greater depth, one finds layered, often mutually supporting agroecological features.

The middle European farmer used long narrow plots. The stated reason was as plowing expediency, ox work more efficiently if they can traverse large areas without having to turn. These long plots, with their high degree of intersystem interface, also helped in controlling the spread of injurious insects and diseases. The flaw lies in being straight, not a problem on a flat surface, an erosion danger when there are hills and dales.

A deeper look reveals the locational vector. Although the accounts are not complete, rye was a commonly grown grain, more so than today. The obvious reason being that rye is tolerant of poor spoils than the more popular wheat. First-hand descriptions from the era understood that plant health is effected by the site and farmers planted grains, rye, wheat, barley, spelt, buckwheat, etc., accordingly.[7]

One has to do some interpolation to discover other mechanisms. Gent (1681) mentions when barnyard fowl, i.e., chickens, ducks, and geese, should not be let into a plot. The inference is that these birds roamed freely through the non-restricted fields, eating ground-dwelling insects.

There would have been more. The loosely constructed farm structures would house barn swallows, chimney swifts, and the like. The ecological niche of these birds lies in their hunting flying insects. There is a one-two punch here, insects scarred up by domestic fowl are open prey to those that hunt on the wing. Of course, wild birds do consume grains. Some crop species deter this threat better that others, this was known and an appropriate species utilized.[8]

The system did incorporate, to a high degree, the available biodynamics. Seriously lacking was the genetic vector. The genotypes planted were not as advanced as current varieties. These would have provided less per plant yield, and may have been more susceptible to

PHOTO 12.1. The Nicaraguan weed with maize case study showing herbicide-killed weeds which provide plot erosion protect.

temperature extremes, plant diseases, and other afflictions. This made agriculture of the period much more chancy.[9]

Substitution

This brief study is from the mountains of central Nicaragua.[10] Maize, the staple crop, is planted on steep hillsides. The fields are not plowed, instead each maize plant is seeded individually.

Rather than have bareground and erosion, farmers allow ample weed populations to exist. The weeds may have a secondary purpose, to harbor predator insects (Photo 12.1).

Since the weeds are not always yield compatible with the crop, the weeds are killed with herbicides prior to planting. The weeds, now dead in place, still serve their anti-erosion function. These reestablish, through in-soil seeds and in-place roots, during and after the cropping season. With this strategy, there are two shortcomings: (1) the need for costly herbicides and (2) the locally recognized danger from spraying toxic chemicals.

With this non-till system, there may be below ground gains. These include nutrient cycling and flourishing micro flora and fauna (provided that the herbicide is not overly intrusive belowground).

The closer a solution is to current practice, the more likely it will gain rapid acceptance. The most immediate course involves substituting the weeds and herbicides with a maize complementary, non-woody, ground-hugging, non-climbing covercrop. There are candidate species, mentioned are canavalia, lablab bean, or velvet bean.[11] A local favorite and therefore first on the consideration list is the cover species *Arachia pintori*.

Management will depend on the dynamics of the cover species. The basic notion is to chop a hole in the cover and seed the maize within the bare spot.

Adoption is less an issue as this is not a radical departure from what is currently done and perennial cover species are utilized regionally. The attraction is the cost and health savings. Other solutions, e.g., woody cover species or progressive vegetative terraces held in place by a nutrient providing tree species, would be looked at only if non-woody covercrops cannot accomplish the task.

Redesign

There is an intriguing case that involves an active search for the best directional vector.[12] The problem in the African case study is that land-use pressures were forcing shorter fallows that provided fewer nutrient and reduced subsequent crop yields. The primary crop was maize, the fallows were long, 10–12 years, and somewhat productive as oil palms provided a secondary income, in the form of palm wine, when destructively harvested at the end of a multi-year fallow cycle.

In an attempt to reduce or eliminate the fallow, four possibilities were tried. The first two involve the biodiversity vector as (1) a covercrop of velvet beans and (2) alley cropping with the tree species *Acacia auriculiformis*. The rotation vector (3) putting groundpeas (peanuts) and maize in sequence. Also possible (4) was the cross-plot in the form of strip cropping where biomass from the strip crop was cut and carried to the neighboring maize strip.

All the alternatives, except the groundpea–maize rotation, met with some success. The failure of the groundpea–maize rotation was blamed on poor in-soil biomass and reduced moisture holding ability.

With the initial cropping success, these options would be fine-tuned. Other factors are looked at. This can be with the anti-insect potential of the cut-and-carry strips, the weed suppression of a maize–velvet bean covercrops, and/ or the insect-disruptive possibilities of the mineral-providing rotations. A serious option, if all goes well, would be to reintroduce oil palms as a parkland over the new agrotechnology. The advantage of keeping the old crops, oil palm included, lies in maintaining some agro-familiarity. The closer the new practice is to the old, the better change that it will gain rapid acceptance.

ENDNOTES

1. The missing agrotechnologies are either not field based (e.g., hydroponics) or are very rare (e.g., floating gardens). If of interest, a floating garden (crops grown on soil-covered floating rafts) has been documented by Rezaul Haq *et al.* (2004).

2. The chemical option for controlling soil erosion can include a top-coat of a binding agent, one that forms upon and holds the soil in place. If the agent is also a biodegradable, time-release fertilizer, this may be less environmentally ominous than some type of synthetic coating. As an added note, potential gaps in practice, such as a soil sealant, come to the fore through a systematic examination of all the options.

3. All in all, this is a fairly speculative look at the interrelationship of various control mechanisms. The idea is to provide some notion as to what might be expected.

4. The various levels of change, not used here, are found in Gliessman (1998, p. 304).

5. A breakdown of the sociocultural factors is in Bradfield (1986), the theory of cultural and outcome in Wojtkowski (2004, p. 273).

6. The alley cropping preconditions, for use in Africa, were extracted from Sanchez (1995), Reynolds (1991), and Carter (1996). Those of live fencing, from Burkina Faso, were derived by Ayuk (1997).

7. Harrison (1775, p. 212) discusses how barley yields and can counter diseases better if site suited. As mentioned, planting different grains also reduces the threats to cropping and, through a better mix of minerals and vitamins, improves diet and health.

8. Again from Harrison (1775), grains with long awns (spikelets) help to throw off grain-eating birds. This is a yet unproven assertion.

9. The state of plant varieties during the medieval times is hard to ascertain. In an example from North America, that from the early 1800s, maize had not acquired frost resistance and measures had to be taken to counter the dangers from a late spring freeze, e.g., Lathrop (1825).

10. The Nicaragua/ maize case was observed by this author.

11. The suggested cover species are from Kass and Somarriba (1999) who have also proposed woody cover species.

12. The African vector search is from Versteeg *et al.* (1998).

13 Multi-Plot Analysis

The landscape is not just a cross-plot vector (as in Chapter 9), i.e., a single agroecosystem providing facilitative services to a nearby plot. Instead the landscape vector has a multi-plot form. A farm is often a mix of yielding or a mix of yielding and non-yielding areas. If well formulated, every plot will ecologically aid nearby plots.

Previous chapters have stressed the plot as the productive center. This is more for presentational convenience. Individual plots can and do ecologically interact and, from this, economic gains are possible. Each level is important, best if the two, the plot and landscape, are co-applied.

For intrafarm dynamics, there are four between-plot scenarios. These have the separate, but neighboring plots, as:

1. economic and ecological independent;
2. economic co-dependent without ecological cross-effects;
3. economic autonomous with ecological cross-effects;
4. plots that are economically and ecological co-conditional.

Complicating this is where the facilitative services are in one direction, radiating outward from an ecologically strong plot, or where such services radiate from each of the plots. The latter, which can involve one or many cross-plot influences, can span an entire farm landscape or involve groups of plots or be co-conditional in two-by-two, three-on-three, or other such co-involvements. Generally, the more the individual plots are ecologically co-involved, the better the potential outcome, with the proviso that the majority of the interactions be positive.

There is more. Farm landscapes are not only productivity stated, cultural norms, and societal values sum across, and become more apparent, when dealing with multiple plots. A single maize plot may not be culturally revealing, the relative positioning and size of many can provide such insight.

How peoples reconcile nature with agriculture is also a cultural expression. Societies put a stamp upon the land through the landscape design chosen.

As suggested above, landscape layout is a very involved topic. At this point, a detailed and revealing economic analysis of all the options remains a distant hope. A one-chapter discourse can only touch upon this often understated agroecological expression, and then only outlining the basic principles and approaches.[1]

PRINCIPAL-MODE AND AUXILIARY SYSTEMS

As presented in Chapter 3, farm landscapes are often a mix of principal-mode and auxiliary systems. The former being those plots that produce agricultural outputs. The latter offers little by way of salables or consumables, but provide facilitative services on a scale larger than found at the plant-on-plant level. A long list of facilitative services can be offered including habitat possibilities (for bug-eating birds, bats, and predator insects), erosion control, and windbreak use.

If a principal-mode system is highly biodiverse and self-supporting in its internal ecology, auxiliary facilitative services are less needed. A prime example is a heavy shade system with natural forest overstory.

If the internal ecology of one system is strong enough, the ecological can spill over to benefit neighboring systems, serving much the same function as an auxiliary systems. A heavy shade system with a natural overstory, because it offers a habitat for beneficial organisms, offers a lot of potential spillover onto a well-designed farm landscapes.

INTERPLOT COORDINATION

Productive areas that are less biodiverse may require the nearby auxiliary plot or plots to supply favorable and output beneficial ecology. Seasonal monocultures are most often in need of this help.

Placing such systems side by side with one or more that are ecologically strong is not a bad start. Farmers often want more and this can be achieved by matching or harmonizing desirable agroecosystem properties (DAPs).

As an example, a mixed landscape of monocultures and agroforests, each with there own unique and non-overlapping DAPs, do balance a landscape in economic orientation, environmental need, and the cross-plot spillover of beneficial, and a few not so beneficial, ecological forces. This type of landscape, one that intersperses rice fields and agroforests, is evident in Southeast Asia. This is also an example where the plots are economic autonomous with ecological cross-effects.

Revenue and Cost Orientation

There can be much variation within and between farms. Some are uniformly intense, others are composed of a small, fertile, well-watered plots amidst larger areas of lesser quality agricultural land. As expected, the better areas are used more intensively, the other plots less so. The resulting in-balance in land-use intensity is easy to observe, but can be troublesome to quantify.[2]

This line of analysis can be revealing and the economic orientation ratio (EOR) can be helpful in this regard. The theory is simple, farmers with large land holdings and few farm resources will seek low average EOR values, those with small farms and plentiful labor resources will look for high average EOR values. More often, a mix of high and low EOR plots achieves the targeted landscape value.

Figure 13.1 shows the revenue-cost breakdown for three farm landscapes. The first, landscape 1, contains four high-input, high-intensity, revenue-oriented plots (the shade areas). Landscape 2 contains three low-input, cost-oriented plots. Landscape 3 offers a mix of plots with three small areas of high intensity, two large plots of low-intensity agriculture.

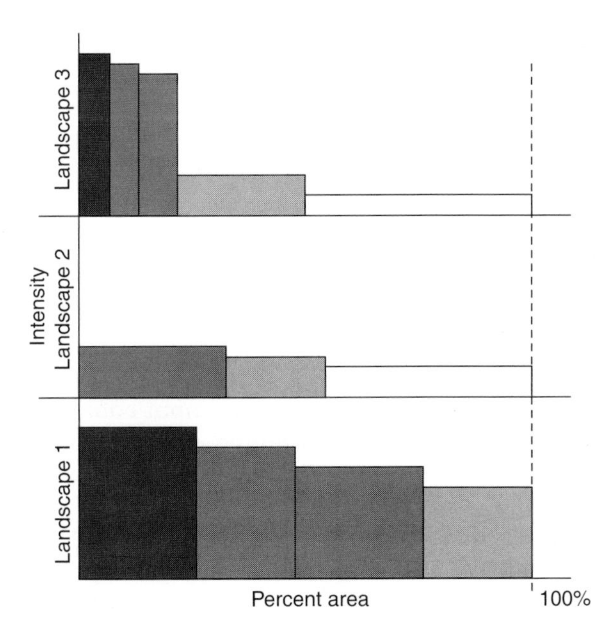

FIGURE 13.1. Three farm landscapes showing different mixes of intense and non-intense plots. The height of each vertical bar represents intensity, the width represents relative area. On the top, a landscape with a mix of intensities (high to low), the middle is a completely low-intensity farm and the bottom has all high-intensity plots.

PHOTO 13.1. A low-intensity agricultural landscape with a few high-intensity plots. Also visible are many agroecological features, riparian buffers included. This photo was taken in southern Chile.

The goal is an average, across-farm EOR value. The average EOR is usually based on labor availability, either average or peak seasonal. Other factors may enter into this. If labor is employed outside the farm, and not as much is internally available, the average EOR declines.

Marginal Analysis

Conventional analysis holds that marginal gains equate across plots, i.e., that the cost of the per-plot inputs, across all plots, provide a like return (the margin gains per unit of input equate). This applies mainly when all plots are revenue oriented, as in landscape 1 (Figures 13.1 and 13.2), less when these are cost oriented or mixed (as with landscape 2 and 3, Figures 13.1 and 13.2).

The idea of equating the marginal gains (as shown in Figure 13.2) still holds, but in a slightly more complex form. The landuser may find that, because of a shortage of farm resources, the cost-oriented options allows overall higher marginal gains than if only revenue-oriented options are utilized. Figure 13.2, landscape 2 has plots 2 and 3 taking the cost-oriented approach. Figure 2.1 shows how these two approaches, cost and revenue orientation, relate to per-plot profitability.

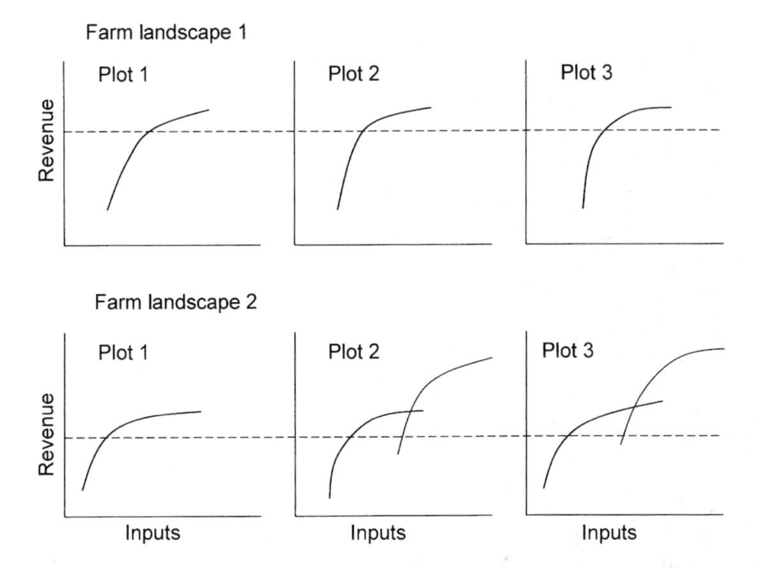

FIGURE 13.2. Equating the marginal gains of conventional (landscape 1) and agroecological (landscape 2) farm. For plots 2 and 3 of landscape 2, the cost-oriented options, rather than those revenue oriented, are equated. Figure 2.1 puts this in economic context.

Economic Spillover

The goal for farmers is to maximize their net return, for the whole farm, not always from the individual plots. Viewed from the economics of a single plot, some practices may seem inefficient. Take the case where maize is sowed at a high density and later thinned or, in desert regions, where wheat is seeded, only producing grain in sporadic years.

Rather than being plot-solitary economic expressions, both these practices are whole farm, i.e., the revenues and costs are shared with other farm activities. In these instances, the maize thinnings and the stunted, non-yielding wheat are fed to animals, another farm profit center.[3] These are the examples of economic co-dependence lacking in ecological spillover.

Photo 13.2 pictures another form of economic spillover. The pasture and the small apple orchard support grazing. Also the cattle can find shelter and some forge with the stand of pine trees. These joint uses, apples and grazing along with the value of the pine stand as source of wood and a shelter for the cattle, complicates landscape analysis. This example (Photo 13.2) is both economically and ecologically co-conditional.

PHOTO 13.2. Economic spillover where a pasture also contains a small apple orchard and a stand of pines.

Ecological Connectivity

If agriculture and nature are to achieve a high degree of harmony, there should be ecological connections across landscapes. Corridors, riparian buffers, and shelterbelts link natural ecologically strong fragments with those that must be periodically restocked with insect-eating insects, earthworms, and the like. In allowing natural flora and fauna move from place to place, their environmental role is not in dispute.[4]

Pitfalls

In trying to maximize ecologically favorable coordination possibilities, there are a few do-nots. Spraying of broad-based, toxic chemicals is the quickest way to destroy insect predator – prey relationships over a wide area, including neighboring farms. Because plant-consuming insects tend to repopulate much quicker than insect-eating types, the long-term aftermath can be very costly. Although barriers can keep insecticides and other agrochemicals from drifting afar, it is better to do without, relying upon the agroecology of the farm for the control.

ECONOMIC MEASURES

To fit within the full landscape, the individual plots must be coordinated or harmonized with other agroecosystems. As a corollary, the full landscape, where each plot is a subsection, requires joint analyses. This is an expansion of those economic measures already proposed.

Landscape LER

The LER, being the basic plot assessment, can be applied to multiple plots and/ or agroecosystems. This is done through the landscape land equivalent ratio (LLER). This is formulated as

$$\text{LLER} = a(\text{LER})_a + b(\text{LER})_b + \cdots + n(\text{LER})_n \quad (13.1)$$

which as $(\text{LER})_a$ through $(\text{LER})_n$ as individual plots or agroecosystems.

The LER is based on the comparative value 1, the LLER is similarly measured. For this, the individual plot areas (a through n) are ratios that sum to 1.0. Take where the four plots makeup 50%, 25%, 15%, and 10% of a farm area, the ratio values are, respectively, 0.50, 0.25, 0.15, and 0.10.

Non-productive auxiliary structures, those that use productive land area for non-productive facilitative purpose are added to the mix, only with zero output (i.e., LER). The gains from having these should be reflected in higher LER valued in one or more of the neighboring plots. The calculation for one of these productively inert plots is demonstrated below:

$$\text{LLER} = 0.25(1.2)_a + 0.25(1.0)_b + 0.4(1.60)_d + 0.1(0)_e = 1.19$$

In this landscape, 10% of the land area is inert. The ensuing gains in productivity are reflected, for example, in plot d.

If one wants to consider price in the LLER, there is the LRVT version. Again, following the set pattern:

$$\text{LRVT} = a(\text{RVT})_a + b(\text{RVT})_b + \cdots + n(\text{RVT})_n \quad (13.2)$$

Cross-plot costs gains are often part of this, therefore there is the landscape equivalent of the cost equivalent ratio (CER) expressed as

$$\text{LCER} = a(\text{CER})_1 + b(\text{CER})_2 + \cdots + n(\text{CER})_n \quad (13.3)$$

Foremost in commercial, and for the vast majority of subsistence farms, there is a requirement that revenues exceed costs. On a plot-by-plot basis, conventional financial analysis determines what is happening. One use of the LLER and the other landscape ratios lie in their value in judging the worth of facilitative plots. With straight financial analysis, there is a danger that these might be overlooked or undervalued.

Landscape EOR

A key analytical detail in an agroecological determination is the degree of land-use intensity. Generally, this is a low–high range where, at the low end is hunting and gathering in natural forests and, at the high end, plants raised in greenhouses. Most of agriculture lies between these two extremes.

The concept of intensity, in some ways, defies quantification, mostly because of the number of variations between the extremes. The EOR, as introduced in Chapter 5, comes to the fore as measure of intensity.

Where farms do not have the inputs to make the full farm a revenue-oriented activity, the EOR often decides which lands are favored with higher inputs, on which lands are put input-prudent agroecosystems, and the resulting land-use intensity mosaic.

Use depends on setting some overall or landscape wide summed EOR. This is weighted as to the area occupied by each measure, revenue or cost, for the EOR. The resulting value can be zero, or a positive or negative figure. This is arrived at through the following landscape economic orientation ratio (LEOR) equation:

$$\text{LEOR} = a(\text{EOR})_a + b(\text{EOR})_b + \cdots + n(\text{EOR})_n \qquad (13.4)$$

For the above, a, b, and n represent the area of each on-farm parcel. These values, a through n, expressed as percent of total area on one farm should, when divided by 100, sum to one. The EOR values for each productively active plot are represented as $(\text{EOR})_a$, $b(\text{EOR})_b$, etc.

Equation 13.4 calculates the LEOR. Although the EOR does not express profitability, it can replace marginal gains as a criterion for allocating resources across many farm plots.

Landscape-Wide Risk

Farmer often seek a situation where some plots are risky, others are safer, and a few will yield no matter what threat nature throws at the farm. This is especially true with subsistence farming or where animal are the key output and these must be provided for no matter what happens.

The basic equation for risk was developed in Chapter 2 can be expanded into landscape version. This expresses landscape risk as an average across a full farm:

$$\text{LRI} = (a\text{RI}_a + b\text{RI}_b + \cdots + n\text{RI}_n)/n \qquad (13.5)$$

where RI_a through RI_n are the risk indices for plots a through n. As with other equations in this chapter, these are weighted as to area.

The weighting factor, variables a through n, sum to one, each one being a proportion derived from the percent of each plot area within the total farm landscape. On contrast to other measures, non-productive plots are risk neutral, i.e., not included in the calculation.

As proposed, for each farm, there will exist some overall risk value or LRI (equation 13.5). Some farmers are risk seekers, some are risk adverse and the LRI sets an overall farm-specific, value.

The LRI is only an indicator in the overall risk picture. More important may be subdividing risk. This can be done by averaging a few high-risk plots, some of medium risk, and a few that yield no matter what happens.

FARM SPATIAL PATTERNS

Within the confines of, or in deference to, climate, topography, ecology, and culture, many landscape layouts are possible. What are utilized at the multi-plot level are the coarse patterns. In contrast to fine patterns (see Figure 4.2), the coarse layouts place agroecosystems, small or large, in close proximity.

The basic coarse layouts are blocks, strips, groups, pivot, and circular. Within each category are numerous subcategories, some involve all seasonal cropping and accommodate rotational needs, others mix seasonal and perennial crops.

These are diagrammed in Figure 13.3. Those on the left can, and often do, incorporate rotational sequence. For those on the right, a well-formulated rotational sequence is integral to the success of these spatial arrangements. As with any design input, these are subject to economic optimization with regard to the crops, cropping areas, rotations, and other variables.

Blocks/ Clusters

Large crop blocks may be less agroecological, offering few cross-plot gains. The opposite occurs when the plots are quite small. This has a comparatively large amount of interplot interface.

If there is any disadvantage, many small plots do not lend well to mechanized agriculture. Also, block systems are not often suited, and gain maximum well-being, in variable topography. Bordered block patterns are pictured in Photo 9.1.

Strips

Strips, with their inherently large amount of interplot interface, offer considerable ecological spillover. Since strips can parallel hillside

FIGURE 13.3. In overview, six interplot spatial layouts. On the left, top to bottom, are a clump, strip, and block designs. On the right, also top to bottom, are of scattered plots, a circular design, and a center pivot system.

elevational gradients, these offer a large degree of topographical flexibility. Since farm machinery can traverse a long swath without making many efficiency-robbing turns, these are well suited to mechanized agriculture.

These come in various forms. Agrobiodiversity strips are crop only. Where season crops must be combined with rotational schemes, varying the planting order in a cross-strip arrangement can serve well. Alternating crop with facilitative strips can be economically rewarding, especially if reenforced by a rotational pattern where the facilitative strip is a temporal fallow. A permanent facilitative strip can be a natural reservation for predator and pollinating insects, insect-eating birds, earthworms, and other service-providing organisms.

Groups

Clumps or groups of a different ecosystem or agroecosystem in an otherwise pure plot may be seen out-of-place. Still, if these intrusions are well formulated and well positioned, they make up for what the larger agroecosystem lacks. Most of the effect comes from a spillover in insect-eating insects and insect-eating birds.

Center Pivot

The block pivots, with their fixed perennial ecosystem in the center, are so named because the surrounding plots temporally revolve, through rotations, around this center. The center features an ecologically strong ecosystem, one that feeds, through spillover, ecological services the outer and ringing plots. The spillover is much the same as provided by plot-internal clumps, the difference lies in the coupling of rotational gains.

Circular

Having a circular arrangement, where the plots are circular bands, such as that of a target, is a possibility. The dynamics can similar to the block pivot, except for the greater amount of what can be all critical interecosystem interface distance. A circular design may have superior wind opposing properties and better resist the spread of harmful insects.

The putting of susceptible crops in the center can be part of a strategy to reduce losses from crop destructive birds. The idea being that shy birds are loath to enter large open spaces to face potential predators. This is the only one control. This is a yet unexplored topic, but, by exploiting the space on all sides of the center, other mechanisms could guard susceptible crops.[5]

Scattered

Scattering is often associated with in-plot crop location placement. As landscape pattern, there are other uses and possibilities. By keeping like crops at great distance, the insects and diseases that attack these particular plants are less effective. Scattering also reduces temperate and other weather related risk. If one plot experiences a severe, yield-harming event, other, far-off plots may not undergo this event to the same injurious degree.

Scattering is found in those societies where the terrain is high variable risk and must be avoided and meaning the climate between individual plots will vary greatly. As a result, scattering is found in the mountainous regions of Africa and South America.[6]

BUFFER SPECIES

An often positive landscape addition can be buffer species. A large amount of interplant interface, e.g., through strips and small blocks, carries the assumption of neighboring plants in neighboring agrosystems have interspecies complementarity. This is not always the case. Farmers may suffer potential losses when the rows close to the system interface are subject to competitive forces. Costs may be increased when sunlight streams under closed-canopies along the open edge. This spurs weed growth.

A well-chosen buffer species can mitigate some of this, defending the system interface from negative spillover while still allowing positive interactions. The upper drawing in Figure 13.4 shows what can happen with both above and belowground competition. In the lower drawing, this is eliminated through the use of a buffer species.

Out of practical necessity, users most choose either the above or belowground effect. For the belowground effect, the most promising are perennial plants with a dense curtain of vertically oriented, fine roots

FIGURE 13.4. Yield and income robbing cross-plot influences, (top drawing) missing after a buffer species is inserted (bottom drawing).

that impede surface spread of other root systems. For the aboveground effect, farmers seek wood perennials with a dense, vertical canopy of a type that gathers horizontal light. For the best result, the canopy would be slight shorter than the tallest bordering species.

Finding species with the full complement of DPCs (i.e., a deep, dense, plunging root form coupled a compact, vertical, canopy at the correct height) has limited what can be achieved. Lacking universal favorites, plants tend to be custom fitted for each application.[7]

COORDINATED ECOLOGY

Presented in Figure 13.3 are some landscape spatial patterns. The textual descriptions suggest some agricultural use scenarios. More conceptional ground must be covered if these are to evolve into something nature friendly.

Blocks and corridors do help in any conservation effort. The farm patterns can be so formulated. This is best if done on a larger regional scale. By adjusting the size of blocks, the landscape can be optimized for various types of fauna, e.g., where smaller birds can exist in smaller areas, especially if well stocked the necessities of life.

Parallel to this is the notion of mimicry. Through patterns and plot content, mimicry attempts to design a landscape that encourages natural flora and fauna. In formulating this approach, it has been observed that paddies and farm ponds have a visible equivalency to natural swamps and ponds and that parkland systems do resemble natural savannas. Despite the speculative attractiveness of this layout option, no natural baseline exists on what might be expected and the similarities with untouched natural ecosystems might be more arbitrary than actual.

For the swamp analogy, the vegetation and aquatic fauna are quite different than that found in nature. This makes these mostly unsuitable for those birds and animals that rely fully upon wet ecosystems. These may have worth in aiding migrating birds, those that seek swamps for an over-night stay with the proviso that crops are not in place when the birds pass or are not subject to damage.

The same holds true with parklands. Superficially, these resemble natural grazing lands, where, e.g., deer, antelope, aurocks, horses, bison, and elephants, were once part of the ancient landscapes of North America, Europe and Asia. These are ecologically distant from cattle only grazing.[8] Still, there are birds and small animals that take to an ersatz mimic.

Mimicry might have worth when harnessing nature for productive purpose. There are the already mentioned examples of birds and bats as an insect-eating agrotechnology. Small mammals can also help, e.g., North American skunks do consume in-soil grubs. A favorable landscape can keep these animals happy and hopefully well fed.

There are questions on how well local flora and fauna are treated through mimicry. Birds and bats may be the main beneficiaries. A landscape with a lot of grazing may end up duplicating, in term of flora, what nature once liked. Where intense cultivation is the norm, mimicry may miss the mark by a large degree.

CULTURAL LANDSCAPES

All landscapes are cultural manifested. Some separate forestry and agriculture, others integrate the two (through agroforestry). The degree of separation or integration, coupled with the suggested agrotechnologies, produces numerous landscape combinations and permutations.[9]

Going further, some groups eschew the notion of landscapes as a series of set, pre-formulated agrotechnologies (the divisional agroecology). Instead, these groups seek more input by heavy modifying agrotechnologies (formulation agroecology) or by placing plant species where they grow best, singly or in mixed groups (placement agroecology), totally discarding the notion of unsullied, ordered agroecosystem.

The proceeding chapters totally rest on the one plot, one agrotechnology model. Some practitioners have may have never looked anywhere but at placement agroecology. Some may have evolved beyond the one plot, one agrotechnology standard into this intuitive and difficult to quantify world.

Placement agroecology is operated byway of guidelines. These are that:

1. each plant species is located where it grows best under the proviso they each does not substantially interfere with a more valuable plant species;
2. the area taken by each plant species correlates with their economic worth as long as the first rule is not violated;
3. groups of plants are placed so that influences (specific interaction zones) overlap, reenforce, or augment across the various climatic and natural threats.

For landscape operated under these guidelines, descriptions are rare and economic results not available. If well designed, there is no reason that these can prove equal in all objectives to a standard one plot, one agrotechnology landscape.

As shown, many factors contribute to cultural agroecology; religion, tradition, and the ecological norms of a society are among the influences.

As what may be the ultimate intangible, landscape cultural motifs oftentimes escape notice. However, they are universal and deep; all farm landscapes have been, and continue to be, visually imprinted by underlying cultural values, priorities, and expressions.

ENDNOTES

1. For an in depth discussion of agroecological landscapes, see Wojtkowski (2004).
2. Discussion as to rating land-use intensity is in Shrair (2000).
3. Tiwari *et al.* (2004) discuss the whole farm, rather than individual plots, as the profit center.
4. Again citing the work of Matlack (1994) and Peterken (1993) with regard to plant colonization and direct connectivity. There is more. The area devoted to wildlife and the results, good or bad, may be economically governed with an efficient optimal solution (Charles, 2002). If developed, this could be part of landscape plan.
5. Circular systems, those based on a circular revolving system of spray irrigation, are fairly common in some regions of North America. As these contain only monocrops without ecological intent, these do not qualify. This could be changed by planting different crops in circular strips. True examples have been recorded in South America, e.g., Stocks (1983).
6. Scattering has been documented in Africa by den Biggelaar (1996) and in South America by Zimmerer (1999).
7. For separating seasonal crops, including crop strips, the perennial grass vetiver, with its dense vertical roots, finds frequent use in tropical regions.
8. Some indication of the ecological differences between parkland and natural savannas is found in Charles (1997).
9. The cultural landscape combinations are from Wojtkowski (2004, p. 273).

14 Agrotechnological Expansions

For a few agrotechnologies, there are more variations than presented. In the interest of conciseness, these are not elaborated upon in previous chapters. However, these represent the full scope of agroecology. In keeping, this chapter expands upon some of the agrotechnologies discussed in earlier chapters.

As with the parent agrotechnologies, it is possible to describe, with some accuracy, the site and user situations where each is best utilized. Beyond this, many come with economic uncertainties.

FEED SYSTEMS

Throughout this text, mentions are made of systems that contain fauna. This can be cattle, horses, llamas, goats, sheep, fowl, or some other domestic, semi-domestic, or a wild animal. Also possible are fish and frogs or useful insects, eatable or otherwise. In the case of domestic fowl, these were once so pervasive in European farm fields that mention was only made when they could cause crop damage.[1] For other farm fauna, inclusion is a specialized undertaking, i.e., to avoid crop damage.

The role of this inclusion is a topic for academic discussion. One view holds that the fauna is the output, the plants being facilitatory with the fauna. There is also the view that the plant is the output, the animal the harvest mechanism. Pastures produce grass, cows eat (i.e., harvest) the grass.

Under the latter view, the forage is the crop, the feed systems are not that unique. Pasture grasses under trees is, in essence, a light or heavy shade system. With the view that the fauna is the output, feed systems take on their own identity.

As with much of agroecology, the best results come when nature is enlisted in the process. For this, the theory is sound. Animals have co-evolved with and integrate into natural ecosystems. These lessons can be applied to agriculture with revenue gains, cost savings, and risk reductions.

Silvopastoral

A large percentage of livestock are raised in pasture settings. This is a low-cost method, attractive in this regard. A simple pasture, one containing grasses or forages, is one application. There are also other forms.

Simple Pastures

Although grass-based pastures dominate as the most common fauna agrotechnology, it is possible to plant pastures of other animal-nutritious forage plants. The cheapest alternative is to let nature take the lead. If all goes well, what will result is short-statured ecosystem upon which the animals will graze. As long as the system is not over-grazed and/ or non-edible plants allowed to dominate, nature, through ecosystem governance, will provide the nutrient and insects dynamics and a host of other ecological tasks. Photo 14.1 depicts a mixed species pasture, one that is totally ecosystem governed.

As the economic standby, the parameters of use are well established. There are revenue-oriented forms, pastures of one-plant species that require effort to keep in their pure state. Mixed forage plants, those that are ecosystem governed, would be cost-oriented. The other consideration takes into account drying and the mid-season loss of forage capacity. Mixed forage would help when this is the norm, the better option might be to add trees.

Trees Over Pastures

There is a class of systems where the trees play no role in feeding the animals, instead these provide an output, either primary or secondary. The grazing animals are another economic return and a separate profit center. Parkland systems do qualify. Outside of this more common design are those with more trees and greater ground-level shade.

The animals can benefit from a sheltered ecosystem. This can improve weight gain and reproduction, especially in hot climates. The shade can keep grasses growing beyond the point where forage in unprotected, direct sunlight would wither.

Deer and other forest-dwelling creatures find these systems to their liking and are a grazing alternative. The advantages are that deer eat the kinds of vegetation that readily grow under a dense tree canopy,

PHOTO 14.1. A mixed species pasture that is ecosystem governed, maintenance free, and hence cost oriented.

even plants shunned by cattle. Because deer and other forest-liking animals do not take well to fencing, this alternative may be best with semi-husbandry (see section 'Semi-husbandry') where the trees cover large areas.

The economics often become less favorable as the amount of shade increases. In this case, the per-area stocking rates decline as the per animal, cost of fencing increases. This necessitates a cost–benefit analysis, comparing the return from trees at different planting densities against the shade produced, amount of forage available, the optimal animal stocking rate, and cost of the fencing.

Forage Trees with Pastures

Grazing animals can and do eat tree forage. One possibility is an alley cropping pasture. For this, the animals feed from both hedgerows

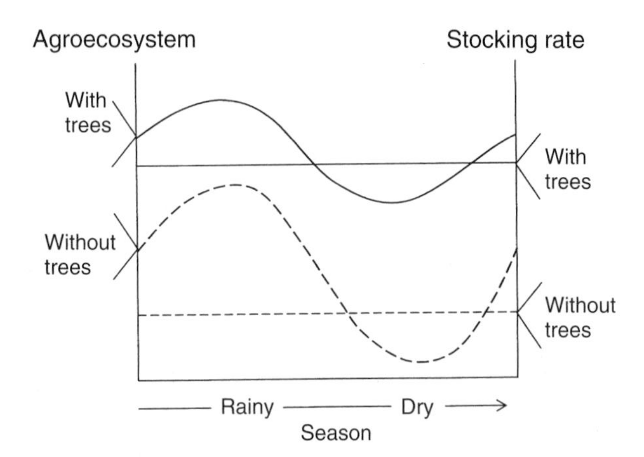

FIGURE 14.1. The difference in stocking rate (animal per area) with and without trees.

and interplanted grasses or forage crops.[2] Gains come as the hedges are often more drought resistant than forage crops. The ideal hedge desirable plant characteristic (DPC) has the animals eating the grasses before starting on the shrubs. This keeps the hedge in reserve should ground-level forage fail or be completely consumed. This is accomplished by planting an edible, but not favored, shrub species.

A more labor involved if this design has taller trees and canopies that are out of reach of the animals. The animals graze on the grasses, when these are in short supply, the trees are pollarded, the pruned leaves are consumed in place of grasses. This frees the landuser from having to pair the grasses and hedge on the basis of relative palatability.

This design is intended for drylands where, as the ground-level forage declines as the dry part of the season progresses, stocking rates are maintained through available tree forage. Under the right climatic conditions, this can be the answer where pastures underproduce. Figure 14.1 visually shows stocking gains from trees and grasses.

Direct grazing of the forage is utilized where labor costs are high and a cost-oriented, sustained pasture is needed. The taller tree version, with pruning, is more revenue oriented. This version might allow for more animal per area, the added costs are absorbed through firewood sales, i.e., from the cut branches.

Forage Trees Alone

Rather than relying upon non-woody, herbaceous plants as the feed source for animals, woody trees or shrubs can provide the needed forage. The woody plants can be grazed directly or these can be part of a cut-and-carry forage bank. Normally, the trees are coppiced to produce the maximum amount of ground-level biomass.

PHOTO 14.2. The Chilean example where tree forage, desert planted, allows year-long grazing.

The advantage is that shrub and trees can be very deep rooted, able to thrive dry or drought condition where other plants succumb. One example is the species *Atriplex mummularia*, grown on the southern reaches of the Atacama desert in Chile. This variation allows year-long grazing where rain is scarce and other forage growth not possible (Photo 14.2).

Forage trees alone are economically compatible with traditional grass-based pastures. The difference is that these are found exclusively in rainfall-starved regions or where rainfall is adequate but highly sporadic.

Aqua-Agriculture

Pond-raised fish are a nice on-farm, low-cost, high-protein food source. They can also be an income source. Trees figure into this as they shade small ponds, keeping the temperature low. The trees can also be the feed source where fruit, nuts, and leaves, falling into the water, can support an applicable fish species (Photo 14.3).

A commonly described example is from China and includes multiple carp species to take advantages of feeding niches (top, bottom, and middle). In Chile, salmon in ponds are fed partially from willow and eucalyptus tree forage.[3]

PHOTO 14.3. A constructed pond in a forest garden where the fish feed naturally off leafs and fruits that fall into the water from the surrounding agroecosystem.

The concept can be expanded. Other aquatic animals, e.g., frogs, can be the center of attention in aqua-agriculture or, where a forest ecosystem surrounds small ponds, aquaforestry.

Aqua-agriculture adds another profit center to the farm. This can be useful when fish and the like sell for high prices. Aqua-systems produce when other systems fail provided that, even during drought events, pond water levels remain high. These systems are useful in humid tropical regions where protein lacks in local diets and is expensive to purchase.

Entomo-Agriculture

Throughout the world, bugs are a food resource with hundreds of species considered edible (Menzel and D'Alvisio, 1998), other insects have non-food uses. Silkworms on mulberry trees are a classic and long-existent example of entomo-agriculture and of a non-food insect product. Although undeveloped in concept, edible insects can be propagated, at low cost, within suitable perennial agroecosystem, to augment diets deficient in protein.

Avian Agriculture

One must not forget birds. Chicken and turkeys are birds that, in their undomesticated state, forage by day on the ground and seek the protection of trees for night roosting. There is no reason that this pattern cannot be reinitiated. The ground cover, as well as the trees, would provide the bulk of the diet for chickens or turkeys.

Avian systems are a cost-oriented alternative to chicken coops and other revenue-oriented systems of fowl production. Use depends on difficult to ascertain tangibles as well as some intangibles. Questions included potential bird losses through straying or from predators. These free-range designs are not used, at least in pure form, when egg output is paramount.

Semi-husbandry

Domestic animals can be allowed to run free. For example, there is the ancient Europe practice of allowing pigs to eat acorns as well as graze on pasture grasses. As mentioned under parkland systems, Chapter 4, this survives as the dehesa system of Portugal and Spain.

It is more than possible to farm, not domestic animals, but with wild animals. These are allowed to exist in a wild state, feed as nature intended, and are hunted or trapped to provide protein in the diet or for outside sale.

Small animals such as ground-dwelling birds (e.g., turkeys) and range of manuals (e.g., dwarf deer, tapirs, and rodents)[4] are hunted, on-farm, to control their numbers. It is only a small step to maintain a residual population, harvesting the surplus. This strategy can be expanded where larger animals, e.g., deer, etc. are more than tolerated, but allowed to consume a portion of the crop. Sections of a farm may be set aside with their preferred forage, crops included.

In contrast to domestic animals, which require barns, fencing, and farm labor, this is a very cost oriented means to obtain meat. In economic terms, this eliminates the sowing-harvest floor. It is possible to view the harvest, in the form of hunting, as a cost-free, recreational activity.[5]

Through mimicry, this concept could be carried to its highest degree. A landscape that duplicates nature could be home to large number of huntable animals. At this stage, a type of semi-agriculture that develops is more hunt only, rather than hunt and gather. Often expressed as ecosystem governed, complex agroecosystems, highly cost oriented are not at all suitable for intense land-use situations. If there is an exception it would be animals raised within an species-rich agroforest (described later in this chapter).

SUPPORT SYSTEMS

Under the biodiversity vector come support systems. These have vine crops grow atop living supports. This mimics nature where vines do the same. As with many attempts to duplicate that found in nature, these systems are generally cost oriented but with a few interesting variations.

Seasonal Support

Seasonal crops, especially vines, require physical support to yield fully. This can be a pairing of productive species, e.g., beans growing upon sorghum or maize.

For a non-productive living support species, many tall plants can fill this role. Most of these will be seasonal, some are alive, others dead, in-place at the time of use.

In the alive category, trees are planted and pruned to keep harvests within reach. The pruning is an off-season activity, the over growth and competition for light from the seasonal vine crop should be enough to keep the tree contained.

In the dead category, non-living stems have advantages, they do not compete for light and water and offer nutrient facilitation in the form of decaying roots and leaves. As a DPC, these plants should have leaves that shed quickly and decay rapidly.

There are three variations of dead supports, those that are (1) planted during a fallow period and killed at the beginning of the cropping season; (2) raised in a strip rotation where one strip is utilized for the initial tree establishment or regrowth, the next contains the dead stems as a vine support; or (3) the woody stem is killed by girdling at the start of the cropping season, the root stock which remains grows new stems between seasons.

There is not enough economic experience to recommend one variation over another. All are revenue oriented, but may be more cost oriented when compared against the use and maintenance of artificial trellises. Two influences may tip the balance in favor of one form over another. The first is labor allocation; worker input for the supporting plants occurring when fewer tasks need tending. The second is a favorable equilibrium in the ecological dynamics; the positives outweighing the negatives.

Perennial Vine Support

Many climbing crops, e.g., grapes, hops, kiwis, vanilla, passion fruit, and black pepper, can be supported by trees. As perennial systems, three variations are possible: (1) vine-over-tree, (2) tree-over-vine, and (3) wine within the canopy.

These come with little economic backing except that derived from successful applications. The deciding variable is often the primary crop (the vine) and where on the support this propagates best.

Vine-Over-Tree

The vine can overrun the tree canopy. This can be with dwarf tree species or one maintained through pruning. Roman agriculture utilized elm or poplar trees for support. This practice continued well into the middle ages.[6] For the modern temperate farmer, the weeping or umbrella elm (*Ulimus glabra*) is short with an umbrella-shaped canopy.

Tree-Over-Vine

It is possible to let the vines grow on the stems of growing trees. Salam *et al.* (1991) reported such as system with black pepper. For this, the vines must be shade resistant or the trees spaced and pruned as to allow ample light to reach the vine. With open canopy species, such as palms, this can be considered as light shade cum support system.

Vine Within a Canopy

It is possible, given light constraints, to have the vine inside a tree canopy. Trees with grapes have been reported and the advantages listed. These include protection against frosts, diseases, and insects. In another example, *Gliricidia sepium* with yams had yield that were double that of unsupported, unaccompanied yams.[7]

Supplementary Support Systems

Vines can be added to tall ecosystem as an after thought. Few examples exist, one being rattan which is planted in low density in established rubber plantations. This does not effect the rubber, but does provide a marketable supplementary output at little cost.

Root Support

Some species have a tendency to lodge or topple under wind and/ or weight stresses. In the classic maize–bean combination, maize has been known to lodge. The beans, with their anchoring roots, can prevent this.

The same can occur with fruit or other treecrops. An accompanying tree or bush can, through interlocking roots, prevent this from happening. Anti-lodging can be strong reason for a second understory species in orchards or treecrop plantations.

RIPARIAN BUFFERS

The simple process of intercepting the soil and nutrients being carried away by surface water has many variations. Their design is similar, polycultural, perennial, with species that are growth active all or most of the year. The importance of year round or season long growth lies in avoiding stream contamination when the defense is dormant. There is also a more the merrier aspect where, instead of unsupported stream bank defenses, the best option is a series of riparian buffers scattered where they can do the most good.

Types

To best clean water runoff, a number of coincident topographical placements is the best course. For this, the design options are (a) simple, stream-side versions, (b) fingered, and (c) arm and hand, and (d) detached types. Described below, these are illustrated in Figure 14.2.

Simple

The most basic of the riparian systems are those located along brooks, streams, ponds, or rivers. As the last defensive structure, these intercept contaminates before they enter the water. These have

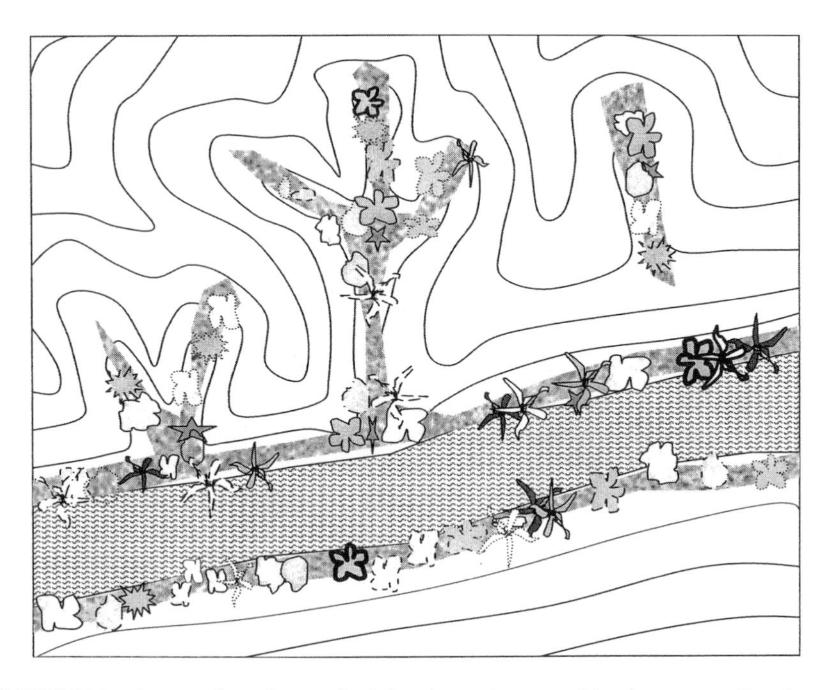

FIGURE 14.2. An overview of watershed showing a stream and land contours. Also shown are four types of riparian buffers: a simple (bottom), fingered (top left), arm and hand (top center), a detached (top right).

secondary purposes, they do shade and cool the water, they can also cleanse or pick up waterborne nutrients that wash by.

There are no set dimensions. They should be wide enough to accomplish the task. If the uphill ecosystem leaks nutrients, these should be wide, if the neighboring ecosystem is ecologically self supporting, then the buffer has less of a function and can be narrow.

Fingered

In regions of slight to severe land relief, dry channels or depressions often lead into and connect with active streams. For most of the year, these remain dry, during periods of high rainfall or runoff, they become active.

A simple riparian buffer next to an active stream will stop a portion of the loose soil and nutrients from entering the water or capture some of that which drift by. This may not be enough. It is better to keep all nutrients, soil particles, and cropping debris out of the water. This is done by buffers that extend up in-feed channels. These are fingers off the simple riparian, meant to fortify these structures in wetter times.

Arm and Hand

In some situations, the normally dry channels are long with many branches. As these might drain a large area, they are inundated when the rains do come. A simple stream-side or short finger buffered may prove inadequate to capture the contaminates. In this case, the channel and some of the branches may be part of the buffer structure. From above, these extensions off the simple buffer will resemble an arm with hand and fingers.

Detached

The land relief in regions can vary greatly. It is always more effect that waterborne soils and nutrients come in as much contact as possible with riparian structures designed to capture these pollutants. Although one continuous system is usually the best, this may not always be possible. There can be situations where these stand alone or are detached from others. These can be located on hillsides where flowing waters converge.

Economics

The problem is that, in the vast majority of the cases, farmers would prefer clean water entering their land and have little regard as to the purity as it leaves. This result is that streams, as they flow across various farms, gather an ever-increasing pollutant concentration.

For cities and towns, it is much cheaper to treat clean water than to remove farm contaminates.[8] Taxes on pollutants or regulations on water purity are the externalities that force on-farm buffer use. Outside of this, it is the vegetative content of the buffers that dictate the economics.

Buffers are perennial where a mix of nutrient-demanding plant species serves best. Within these rather loose parameters, farmers are free to insert almost any content.

Permanent shrub gardens, with a mix of fruiting tree and shrubs, are one containment option. These differ from the shrub gardens discussed later in this chapter in that cost oriented, rather than revenue oriented, formulations are preferred. Buffers may produce fruits and the like, but as a decoy crop, to keep birds from feasting upon more valuable offerings. Buffers may be utilized as cut-and-carry feed systems. Allowing grazing animals to feed directly can reduce the pollutant-control effectiveness of riparian buffers. The most common of options is to comply with whatever rules or laws force installation in the most cost-effective manner. The latter generally offer little or no financial return.

COMPLEX AGROECOSYSTEMS

In Chapter 4, complex agroecosystems were introduced. These are observed as agrobiodiverse, continual agroforests that contain many productive species intermixed in a way that duplicates the natural forest. Although fronted by the agroforest, this is not the only viable complex agroecosystem.

Pastures of mix grasses and other non-wood perennial species also fall under this heading. Agroforests and mixed species pastures are found in the temperate and tropical regions. Because both are highly productive, cost efficient, and integrated well with nature, these represent a zenith in the nature/ agricultural interface.

The ideal is to reduce costs by allowing natural biodynamics a free hand. Except for minor maintenance, the principle cost is the harvest. For the mixed species pasture, hay is the output, the harvest of which is the main expenditure. Photo 14.4 shows an end-of-season cost.

Another expression is the complex, continual garden. This garden is composed of seasonal species with many sub-rotations. Harvests are staggered and, when one crop is harvested, it is replaced by a like or unlike species. To be continual, these are only found in tropical regions. In contrast pastures and agroforests, complex, continual gardens are mostly revenue oriented.

PHOTO 14.4. The only input in this ecosystem governed, biocomplex pasture is the hay harvest.

Natural Compatibility

Studies have shown that in all aspects of sustainability and beyond, few systems compare to the agroforest. Even natural ecosystem fall short is some desirable gauges of what an ecosystem should accomplish.

Variations

The agroforest is based on biodensity, biodiversity, biodisarray and, as continual or perennial agroecosystem, bioduration. These parameters are discussed in Chapter 4. As the name implies, trees are integral in these systems. The pasture version is similar ecologically, only lacking trees as the major component.

Despite the biodisarray, pattern are found and are evident. As mentioned in Chapter 4, user put species where they will do the best. They also pair species with known complementarity and keep competing plants apart.[9] The result may be a kind of semi-pattern, not random but still disarrayed. Figure 14.3 hints at what can be done in this regard, detailed economics for these pattern variations are not available.

Homegardens

As a very common system in the humid tropics, homegardens are often located near or around households (as in Photo 14.5). So much so

FIGURE 14.3. An overview of two agroforest layouts, each exhibiting a different form of patterned disarray. On the top, the upperstory is ordered, the understory disarrayed, evenly scattered and evenly spaced. The bottom version is of a disarrayed, clumped pattern.

as a canopy gap is formed above the actual house. A unique feature is that the surrounding gardens are often fertilized with organic household waste.

As to the types of crops, some are devoted to annual crops, others are almost totally treecrops. Those plants useful to households are generally

PHOTO 14.5. A house located in the middle of homegarden.

in evidence, especially spices and medicinal species. Domestic fauna can be part of the picture, with pigs and goats, chicken and ducks. Ponds and fish can also be included in a homegarden.[10]

These are not entirely a tropical agroecosystem. Temperate versions exist in southern Chile. Some French potagers can also be inclusive in this category. In addition to being part of the eatable landscape, providing food and medicinal herbs, homegardens often contain many flowering plants.

Shrub Gardens

Shrub gardens are midway between a homegarden with few trees and the high forest version. As such these consist of productive woody shrubs (e.g., coffee, cocoa, pigeon pea, and papaya), short-stature fruit trees (e.g., citrus species, rose apple, and pandanus), longer-duration non-woody plants (e.g., pineapple, cassava, and bananas), and vines (e.g., yams and black pepper). Some are permanent, others are spatially transitional (will mature into a full size agroforest) or temporal transitional (will be replaced as these are more of a productive fallow).

The permanent systems are generally high-output, revenue-oriented, based around perennial, market-oriented species such as coffee and cocoa. Spatially transitional shrub gardens are also permanent, but are not tall. This is because they are located between vegetable gardens and taller forest gardens. The temporally transitional shrub gardens are a part of a cropping cycle and generally contain short-duration plants, e.g., cassava, chili peppers, papayas, bananas, and cashews.

Forest Gardens

Forest gardens, also called agricultural forests, are fully forested ecosystems except that they contain all or a large percentage of productive, often fruiting trees.[11] If one were to formulate a species-complex temperate agroforest, one would interplant various varieties of apples, pears, peaches, plums, mulberries, chestnuts, walnuts, and butternuts. The intent is to visually, and ecologically, duplicate the natural forest.

Of course, bushes and vines are encountered in forests and certainly currents, gooseberries, raspberries, and blueberries are proper shrub additions. As for vines, grapes and kiwi fruits would serve well. Seasonal species would not be out of place and shade resistant crops, such as squash, beets, beans, etc., can be included as the agroforest understory. What occurs is a cornucopia of outputs at little cost.[12]

As with the homegarden, birds and animals, natural or domestic, are a possible inclusion. Ponds containing plant-eating fish (as in Photo 14.3) also increase the in-forest protein harvest and the food value of a forest garden.

Intense forest gardens can contain up to and over 200 productive species, more if unnoticed weeds are counted. A breakdown might be along the following lines: (a) a dominant population of a few (two to four) productive species that form 25% to over 75% of the total plant total; (b) commonly incorporated plants that makeup 5–25% of the plant total; and (c) a large number (percent) of trace species which individually makeup less than 1% or 2% of the total plant population.[13]

In an undeveloped form, agroforests are enriched natural forests. This can be managed as natural forests where useful species are encouraged. Through a more active involvement, useful species are actively planted. As the forest shifts, the end goal is the more intense, revenue-oriented version.

Forest farming is a version of the forest garden with crops as the understory. Crops can be in strips within the forest or raised below the canopy. Examples of below-canopy crops are mushrooms, berries, and decorative plants along with specialty fruits and berries, e.g., chokecherries, elderberries, sand cherry, hawthorn, and America plum.[14] Where open strips are utilized, any commercial crop that seeks vertical light is a possibility. As with treerow alley cropping (see Chapter 4), the row or strip direction, usually north–south, may be an important design attribute.

Management

The unique dynamics, as least when compared with conventional agriculture, dictate unconventional management. The following management, or rule-of-thumb, guidelines were suggested by Wojtkowski (1993). These are:

(a) if a plant output is needed and the plant is producing well, leave it; if not, alter the competitive environment;
(b) if its production is not needed, neglect it;
(c) if it is negatively influencing a more desirable output, prune it;
(d) if space exists and essential resources are underutilized, as determined by the amount of light striking the bare ground, plant or let something grow.

For the above, seldom, if ever, is anything removed. Unwanted plants do add to the overall biodiversity and are generally kept. What may happen is that weeds are suppressed, mainly through shade or combination of pruning allowing belowground interactions to take their competitive toll. These unwanted plants, placed at a disadvantage, may eventually succumb, as in natural ecosystems, to the forces of nature.

Economics

Despite their popularity and the many articles on agroforests, economic analysis is mostly missing. It is widely assumed that these rank high in overall productivity, as measured by land equivalent ratio (LER). The difficulty in LER valuations and with financial evaluations is that the outputs are many, the per-unit output is small, the non-harvest option is the norm, and yields for many species may be seasonally sporadic. The latter occurs when plants do not produce each season. This makes it difficult, if not impossible, to gather meaningful yield data.

On the other side of the profit equation, agroforest costs are hard to track. Much of the labor is casual, a worker or household member may enter the agroforests and, while transiting or harvesting, take a few minutes to perform some needed task. Cumulatively, such casual inputs are difficult to fully quantify.

The missing economic information may be one reason why agroforests are of less commercial interest. Still, examples abound, their clear-cut, on-farm success, rather than the missing economic data, insures their popularity.

OTHER DIMENSIONS

Agroecology does exist outside the confines of agriculture. For one, silviculture is fully governable by the same principles and concepts.[15] There is yet another dimension, agroecology as defined in terms of the non-agricultural, non-silvicultural, ecological services provided to people.

Riparian buffers exist outside farm landscapes. Windbreaks, sometimes as snow fences, are also ex-farm. These are straightforward, people-oriented, ecological services that fall under an agroecology heading. These are use determined through cost–benefit economics. These set a boundary on agroecology.

Crossing the line, the ecological services offered under conservation ecology are not directly people oriented. Instead, they help nature recover from the degradations inflicted by mankind. As there is no expectation of any direct financial return, there are outside of cost–benefit analysis. The only expectation is that the effort be done at a reasonable cost.

ENDNOTES

1. The lack of mention, except where exclusion is necessary, of free-ranging domestic fowl (geese, chickens, and ducks) is from Gent (1681).

2. There are many citations on situations domestic animals consume tree forage. One such discussion is in Bryan (1999).

3. A description of carp ponds in China can be found in (Gongfu, 1982). This author has observed pond-raised salmon, under trees, in Chile.

4. Dwarf deer are found in West Africa and a different type (the pudu) in southern Chile.

5. Semi-husbandry is presented in Linares (1976), the types of plantings by Green (1908, p. 52) and Robinson (2005).

6. The Roman example is from Lelle and Gold (1994), that from the middle ages is a first person account (Chaucer, 1382, line 177).

7. The trees with grapes report is from FAO (1994). The doubling of yam yields is in Budelman (1990).

8. There is an agroecological way to clean flowing water. Soil covered rafts planted with vetiver are said to function well. The economics, as compared with formal water treatment, are uncertain and may depend on various factors. In their favor, vetiver rafts are low technology enough for use in many locals.

9. Pattern of agroforests are recounted in Gillespie *et al.* (1993), Aumeeruddy and Sansonnens (1994), Michon and Mary (1994), Gamero *et al.* (1996), and Méndez *et al.* (2001).

10. Among the numerous authors providing insight on homegardens are Jensen (1993), Gillespie *et al.* (1993), Aumeeruddy and Sansonnens (1994), Kumer and Nair (2004), Michon and Mary (1994), and Ali (2005).

11. A discussion of the forest garden is in Belcher *et al.* (2005).

12. The low-cost figure of agroforests are suggested by Cooper *et al.* (1996), Soemarwoto (1987), and Torquebiau (1992). This may not be universal, e.g., Ali (2005), but such differences may be regional.

13. The species breakdown is suggested in Raynor (1992), Babu *et al.* (1992), and Salafsky (1994).

14. The below-canopy crops are from Dix *et al.* (1997) and Josiah *et al.* (2004).

15. Silviculture as an agroecological science is outlined in Wojtkowski (2006b).

15 Analytical Refinements

Through the earlier chapters, a number of economic concepts have been presented. In some cases, these are in introductory form. There is reason for this. The initial emphasis is not on fully developing each methodology, but on showing how each fits within an overall analytical flow. Where advantageous, this chapter analytically expands on this.

Economics is not, or should it be, the sole criteria as to what a farm ultimately looks like. There are influences that fall outside the overlapping spheres of agrobiodynamics and agricultural economics. For the sake of completeness, these influences are also described here and their role explained.

Once all the pieces have been examined, the larger problem is how to sort through that presented and reach meaningful conclusions. Bioeconomic modeling can be a substantive application of principles and concepts or it can be a means to guide the thought process; helping to consolidate and concentrate on the most promising avenues. In its integral form, a bioeconomic model can also be an unmatched analytical tool; concluding the previously described agrobiodynamic and economic pieces.

LER VARIATIONS

Searching for the best performing (i.e., the highest land equivalent ratio) crop combinations can involve some detailed research. Where, in order to bring about an improvement, a breakdown of the biology is often warranted. This can necessitate cleaving a design or agrosystem into separate ecological events.

Row-Based LER

A common occurrence is finding the ideal planting density and, for this, a row-by-row breakdown is insightful. When rows contour and climb a hill, it can be useful to find at what point (elevation) when it is best to change the variety, the crop, or the treatment. Standard research designs progressively vary the internal density of each row. From this, an optimal spacing must be determined.

The row-by-row LER is

$$\mathrm{LER} = ((Y_{ra} r_a)/Y_a) + ((Y_{rb} r_b)/Y_b) \tag{15.1}$$

In the above, Y_{ra} and Y_{rb} are the average yields for one row of species a and b, respectively. Additionally, r_a and r_b are the number of rows inclusive in the sample. As is standard with LER analysis, Y_a and Y_b represent the monoculture yields, but for the same number of rows as the numerator sample.

Event-Enumerated LER

Improvement may come through an understanding of underlying principles or mechanisms. In many cases, it is possible to separate the underlying happenings. For example, if one wishes to see if root interactions are negative, root barriers between unlike species will eliminate, at least in the short term, this interaction. For this, the LER value is based on a single happening, e.g., the economics of light interception.

A mechanism or event-enumerated LER can take the following form. This is a facilitative situation, species b influencing species a, not the reverse:

$$\begin{aligned}\mathrm{LER}_{(\mathrm{event})} = 1 &+ ((Y_{ab}/Y_a)_1 - 1) + ((Y_{ab}/Y_a)_2 - 1) \\ &+ ((Y_{ab}/Y_a)_3 - 1) + (Y_{ba}/Y_b) + c\end{aligned} \tag{15.2}$$

In this reformulation, three events or effects (1–3) are itemize where

$$1 + ((Y_{ab}/Y_a)_1 - 1) + ((Y_{ab}/Y_a)_2 - 1) + ((Y_{ab}/Y_a)_3 - 1)$$

substitutes for (Y_{ab}/Y_a). For exactness, the latter constant, c, reconciles the error factor, hopefully small, between the ordinary LER and the event-enumerated version, i.e.,

$$\mathrm{LER} - \mathrm{LER}_{(\mathrm{event})} = c \tag{15.3}$$

A three-event breakdown might include an analysis of roots, shade, and shed biomass. Event 1, i.e., $(Y_{ab}/Y_a)_1$ can be determined by placing artificial shade, e.g., a sunscreen, above species a, determining shade on yield. Event 2 has root barrier between the two species, without any on-ground biomass, the upshot of root-on-root competition can be isolated. Event 3 is where species b is absent, but the fallen biomass from species b is collected, carried, and placed next to species a.

For agrobiodiversity systems, where the cross-species influences go both ways, Y_{ba}/Y_b would be similarly event enumerated. Again, this can be into any number of effects.

Not forgetting costs, a similar event-driven equation is

$$\mathrm{CER}_{(\mathrm{event})} = 1 + ((C_a/C_{ab})_1 - 1) + ((C_a/C_{ab})_2 - 1) + ((C_a/C_{ab})_3 - 1) \quad (15.4)$$

Three events, 1–3, represented a redesign; one with a number of changes. If everything is the same, except for a single substitution, the comparison is simpler, i.e.,

$$\mathrm{CER}_{(\mathrm{event})} = C_{a1}/C_{a2} \quad (15.5)$$

For this, C_{a1} is the old procedure, C_{a2} the new. The new procedure could be added microbes, a change in tillage, or some other substitution.

In a comparison of ridge tilling (C_{a1}) against no-till (C_{a2}), the respective costs are \$199 and \$143.[1] The result is 1.39, indicating that the no-tilling method is a major cost variable.

LER Pitfalls

Issues arise when dealing with LER, most are denominator related.[2] The clear favorite is to employ an optimized monoculture; one shorn of random happenings. If insects appear in scattered seasons, this might require that experimental plots, those utilized to determine the monocultural base, be liberally doused with insecticides to provide a risk-free base figure. If insect losses are the seasonal norm, then the deprivations of these bugs would be included in the base figure and the hoped for anti-insects properties of an intercrop would be reflected in the LER. In either case, insects or other happenings, random or not, might be inclusive in an event-driven LER analysis. The LER is characterized by its flexibility; not that much harm is done with sub-optimal base.

CONTINUUMS

Rather than viewing agroecology as a series of discrete agroecosystems, agroecology is better viewed as a range or spectrum of designs, each a slight variation of another. Agroecology is fraught with such continuums. These span most all-active variables. The production possibilities curve (PPC) is an example of a continuum, that in Figure 4.4 represents differences in plant populations.

In addition to density, the amount of biodiversity can vary from high to low. One can modify spacing, the spatial pattern, and/ or make changes in the timing of the crops. The list of included variables is quite large.

Although seemly an innocuous, the possible tradeoffs, and the form these take, are an integral part of the understanding and evaluation the outcome. In short, these can be a study unto themselves.

Agroecology offers many one-on-one tradeoffs. The space freed through the removal of one plant species can be filled by another. The yield reduction, due to competition for essential resource by a second species, is supplanted by the yields from the second plant species.

In a surprising number of cases, one–one tradeoff continuums have a third dimension, that of the best-of-all worlds. These alternative formulations are not always useful, but are interesting.

In going from revenue to cost orientation, the tradeoff can be one-for-one, for each unit of input taken away, one unit of output is lost. This may not be the case, somewhere along this continuum subtracting input units may not result in output losses. There are even situations where yields increase as inputs decrease. The latter are the most profitable and sought after design solutions.

BEYOND THE PPC

The PPC represents a continuum of options. Either in their single season form (Figure 4.4) or across time (Figure 6.1), these foretell which co-species, spatial pattern and joint planting densities give the best LER.

A more complete picture comes in adding costs and, for this, the PPC and cost equivalent ratio (CER) curve can be superimposed in the same graphic. The better option is to present the $RVT_{(CER)}$ (relative value total) in curve form; a derivation similar to the PPC or CER curve as already demonstrated.

Curves based on the $RVT_{(CER)}$ have one slim advantage over individual profit points, the ability to interpolate and arrive at conclusions outside an immediate data set. This allows for a higher degree of design refinement.

For plot or even landscape analysis, this may be as far as ratios will go without stretching these beyond economic meaning. Past this stage, it may be better to transition into direct, ratio-free, financial analysis.

SPATIAL CONCERNS

In numerous cases throughout this text, spacial pattern has been shown as a critical determinate. After species, this is the second most addressed variable. A lot rides on the pattern chosen. The LER and/ or the CER can be aided by a good selection or much can be lost through a bad one. This is knotty question when biodiversity levels are high and knowledge low. Spatial disarray shortcuts the question.

Spatial Theory

There is a small body of thought on spatial patterns. The importance is in finding the correct spatial pattern given the species, soil, weather, and other conditions.

Figure 13.2 shows six bicultural spatial patterns. These can be subdivided into fine and coarse. Fine patterns have each plant species touching an unlike plant species. This maximizes the amount of interspecies interface, the resulting biodynamic gains, and the potential LER. Tight interspecies distances can also be a mechanism for cost control.

Course patterns limit the amount of interspecies interface. This is useful if rows or groups of facilitative plants are needed to fully support one productive species, as in Figure 5.3, or when unlike species are competitive if densely interplanted but, through spill over, confer one or more cross-benefits. Use depends on the LER potential for the species mix (two or more).

An informal rule of agroecology holds that those species combinations with highest LER potential should utilize a fine pattern, those of lesser potential may find a coarse pattern more accommodating. This rule comes with caveats but, in general, offers a good starting point.

If two species are competitive and have a positive spillover, there is always the option of a buffer species (see Figure 13.4). As internal plot boundaries, these keep two competitive species from direct contact while allowing for ecological spillover.

The division of essential resources within an agrotechnology is also part of spatial theory. A taller plant can be sunlight blocking, either for the good or bad, for a smaller species. If direct sunlight must reach the understory for a span of time, a coarse pattern is better. Treerow alley cropping differentiates from shade systems because the alley allows the understory access to direct sunlight (as in Figure 4.6). The timing of the direct light is a function of the strip width for the understory and spatial orientation of the strip.

Under normal circumstances, treerows are situated in a north–south direction. This allows vertical, midday light to reach the between crops. This might not be the best layout, if crop experiences the best growth in the mornings, the row pattern should be oriented so that the crop receives maximum light at this time. Take the situation where morning dew is heavy upon the ground, the crop, less subject to drying, might better utilize maximum light in the early morning. In the afternoon, when all is dry, excessive light might prove harmful. In the northern hemisphere, this strongly suggests a southeast to northwest strip orientation will offer the highest possible LER.

Light may not be the only row or orientation influence. Winds can be especially damaging if channeled down long rows. Stopping the wind tunnel effect may require a different placement or spatial pattern.

Of course, other variables intercede, e.g., if machine harvested, a coarse pattern may be best. This is especially true if the component crops produce an unlike output or if harvests occur at different times.

Disarray

As mentioned, biodisarray, i.e., the lack spatial order, is a possibility. This can be undertaken for cost purpose. For wheat, rice, and like grains, it is cheaper to broadcast the grains rather to row drill individual seeds. As long as the per-area spread is within set top and bottom limits, no economic harm is done.

Disarray outside of monocropping is a different study. Spatial disarray with complex polycultures can sidestep the messy situation that occurs in not understanding on the specifics of plant-on-plant competition. To achieve acceptable LER values, spatial order necessitates that certain species be located next to other species and that these have strong interspecies complementarity. Often these relationships are not known and an LER-successful, ordered interplanting, one based fully on imperfect knowledge, would be chancy at best, i.e., be far from optimal pairings and densities (for each in the many species).

Rather than gamble on an ordered system, one with many component species and without base knowledge, it may be better to utilize biodisarray. With no order, there are no set neighbors, no predetermined interplant spacings, some pairings will exhibit success, others not so. In aggregate, a level of achievement and a base LER should be higher than with a poorly selected, co-planted species utilizing systematized spacing.

There is another facet of disarray, that of the fluctuations in seasonal weather patterns. Even if best polycultural order is known, including the best pairings and spacings, there is no guarantee that nature will cooperate. Pairing and patterns are predicated on essential resources being above some lower limit. Good plant pairings can turn competitive if certain levels of resources are not available.

Competitiveness could be magnified in biocomplex agroecosystems. A species mix that returns a highly favorable LER when conditions are auspicious, can be decidedly bad when drought or some other negative event befalls. Again, the solution is biodisarray.

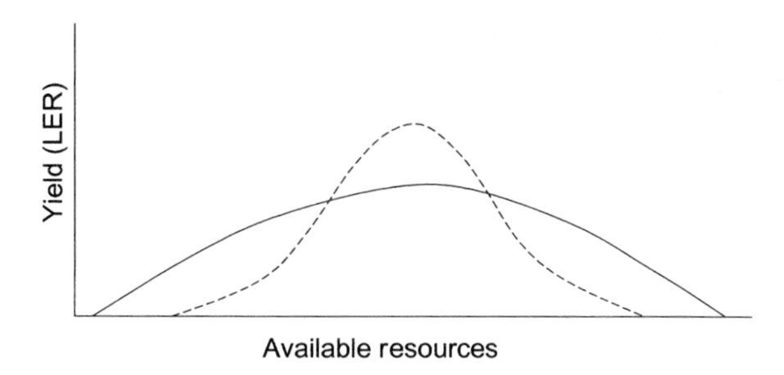

FIGURE 15.1. The effect of disarray and non-disarray as available resources change. Although yields can be higher with an ordered system (dotted curve), this requires optimal resources. In contrast, disarray is better if resources are highly variable.

Disarray, with its mix of plant pairing and spacings, tends to level the highs, but offers greater returns at the extremes. The system will not be as successful when growing conditions are good, but will return something if weather patterns are less promising. Figure 15.1 shows a comparison, what happens with the LER when ordered and unordered polycultures confront normal and the extremes of weather.

Patterned Disarray

Disarray may hint at, but does not denote, random plantings. Users customarily place species, those with known complementarity, in close proximity. With knowledge, those that are competitive are consistently located far from each other. Randomness only occurs with new additions, those that arrive without advanced undertstanding.

Within this context, patterns do occur. It is not unusual to find groupings, species that coexist well are planted close, those that are slightly more competitive are further apart, those that are highly competitive are at a greater distance. This results in small intrasystem groupings and with some, often insignificant, loss in homogeneity. Many homegardens are internally patterned in this way.

Semi-disarray is also possible. One or two species, usually the tallest, may be placed in some symmetrical configuration, the others not. Patterns may be discernible, but not precise (as in Photo 15.1). Such variations constitute spatial options. In current practice, a lack of mastery of the pros and cons on their exact positioning limits use.

TEMPORAL INCOME GAPS

Among the application problems, especially when changeovers occurs, are yearly gaps in income. These income gaps can occur when

PHOTO 15.1. A plot showing semi-disarray where the spacial pattern is discernible, but not precise.

seasonal rotational sequences contain low value crops or must be interrupted by a nutrient gathering fallow. Income gaps can be more pronounced when permanently shifting from seasonal cropping to a woody agroecosystem (orchards, treecrop, or forestry plantations).

For a seasonal gap, the primary counter is to divide the farm so the affect of the low or no income period is lessened. For example, if the low income crop or the fallow occurs once every 4 years, then the plots are arranged so that one quarter of the area is planted in the low income or in a fallow in any given season. This spatial/ temporal arrangement avoids a seasonal income gap. Since the purpose for the changeover is higher net income, no overall income loss is expected, i.e., this should be inclusive in the economic feasibility calculation.

A changeover income gap can be more severe if, in the first few years, the trees are not yet mature enough to provide full or any income. Economically, this involves going from a lower income to a higher income plateau with a transitional income loss (as shown in the solid line, Figure 6.1). With fruit trees and the like, crops interplanted with the trees can reduce or eliminate this gap, i.e., as the crop yields decline, fruit yields proportionally increase.

Other scenarios are not so forgiving. If the end goal is a facilitative shade system or wood-producing tree plantation, the gap may persist but, when the temporal plan is changed, in a reduced form (as in the dotted line, Figure 6.1). The problem is that it can take years for a more profitable woody perennial to reach maturity. In the case of on-farm forestry trees, it may be 25 years until the first harvest.[3]

The most potent, gap-reducing weapon is to convert sections of the farm to the newer, hopefully more profitable system. As these come economically on line, gradually adding more area until the entire farm is transformed. This process often spans decades. Usually the worst yielding agricultural land is the first converted.

Another gap-reducing measure is the simple taungya. This involves crops interraised with trees in the initial years (as discussed later this chapter).

The planting method can also have a large role. Instead of seeds or seedling, the trees are transplanted from the tree nursery in a more mature form, e.g., as striplings (a tree of 2–3meters height striped of lower branches) or long stems (a live pole of 4–5 meters height planted without roots or branches).[4] The mature-pole planting methods greatly hurries the establishment process by as much as 10 years, reducing the temporal income gap.

OVERALL RISK

The most understated aspect of agriculture and, by extension, agroecology, may be overall on-farm risk. This is more than landscape-wide. As an expansion of the Landscape Risk Index concept, other factors enter into this.

Where available, farmers can mitigate risk through crop failure insurance. In diversified economies, outside employment offers an escape. Where such ex-farm, anti-risk alternatives exist and farmers do not face starvation when normal crops fail, potential risk does not attract the attention it deserves.

Where farmers cannot rely upon outside help and the consequences of crop failure is dire, anti-famine counters may rise to the fore as a key decision variable. What complicates this line of analysis is that the anti-risk measures are many and varied.

For some, large herds can be sold or directly consumed to bridge a shortfall or failure in the staple crop. This may be an unacknowledged rationale for overly large herds. To bridge shortfalls, others may seek food from natural sources. If near the sea or other bodies of water, fish can be caught. Wild animals, as well as nuts and fruits, have a long, often overlooked, history as a backup food source.[5]

Others seek cropping solutions. For subsistence farmers, a few low-risk plots avoid starvation during long periods of unfavorable weather. A few plots that yield, no matter what happens weather-wise, allows farmers to maintain other crops that are comparatively risky, but with a higher profit potential.

Where crops or livestock are highly risky and seasonal rainfall is less than adequate, the counter may involve a move to where life can continue. This may be the essential anti-famine strategy for nomadic desert societies.

Another complication, this analysis is comparative. In high rainfall West Africa, farmers plant cassava in the event that water-demanding upland rice yields poorly. In drier regions where cassava is the staple crop, drought-resistant treecrops are often the backup food source.

The implications loom large. Identifying the anti-adversity or anti-famine countermeasures and, through this, keeping the landscape risk index constant may placate farmers worried about going out on the preverbal limb.

Take the case, in higher risk climates, where rising revenue orientation increases the risk index. The introduction of a secondary, anti-risk crop, especially one that is very cost-oriented, may alleviate famine or penury anxiety. Eliminating or lessening this anxiety may allow greater headway in introducing new crops and/ or cropping methods.[6]

MULTI-PARTICIPANT AGROECOSYSTEMS

For the sake of completeness, multi-participant agroecosystems do find use and deserve mention. Under the multi-output scenario, different groups and/ or individuals assume responsibility for the various crop components. Taking the classic maize–bean intercrop, in theory it is possible for one farmer to grow maize, another beans, together on one intercropped plot. Coordination difficulties generally make this unfeasible.

The commonly encountered multi-participant system is with the simple, tree-plantation taungya. The farmer raises the crops in the first year, the forester plants the trees. As the trees grow, the farmer relinquishes farming rights in the now established plantation and resumes activities in another first-year tree plantation.

There are a lot of advantages to this. When each operates independently, each must prepare the land, weed and fertilize their respective crop. The sharing of these costs makes sense.

Other gains lie in the knowledge required. Each may understand one of the crops, know little about the other. Pooling their knowledge avoids cropping mistakes and introduces a higher level of efficiency.

Another gain is in specialized equipment. Farm machinery can be expensive to operate when each party, operating alone, utilizes it over a comparatively small area. Through a multi-participant agreement, the equipment covers more ground each season, the cost being amortized across greater areas and larger harvests.

Other than taungyas, the multi-participant ecosystem can apply across rotations. Farmers specialize in one crop species, neighboring farmers another, and so on. With rotations, each farmer has plots of their forte crop, either on their own farm or on that of a neighbor. Seasonally switching farms might be considered a drawback, but again, a specialization in knowledge, in machinery owned, and the better operating and expense structure for that machinery are clear economic pluses.

The cost savings are apparent, but this does not seem a strong enough reason for widespread application. The main barrier lies in the accepting the required cooperation and compromises. These must insure that (a) there are few, if any conflicts; (b) the each participant understands their role; and (c) the agroecosystem design takes the needs of each into account.

To bring home this point, most practical multi-participant arrangements are through simple taungyas, rather than more complex agroecosystems. Still, going down the design road, i.e., finding crop or rotational combinations that can be cooperated, could make multi-participation more accepted.[7]

AGROECOLOGICAL MANDATES

The agroecological mandates require going beyond profit and loss. Commonly, risk is not part of formal profitability analysis, but is part of subsequent agroecosystem yes-or-no decision process.

The process continues, adding other decision variables or determination criteria. Inclusive in pre- or post-profit analysis are culture or societal values as well as preservation of native flora and fauna. Beyond this comes clean water, area climate, esthetics, and so on. How these mandates are handled is an unresolved issue.

As stated in the first chapter, there are degrees of precision. It is a lot nicer if most of these mandates are tangible and quantifiable. To be quantifiable, three obstacles must be overcome, there must be a:

1. base number from which to analogize,
2. ranking scale based on a quantifiable underlying premise,
3. value range that is both user intuitive and cross-comparable.

The LER is formidable index in that it has both a base point (1.0) and an intuitive and cross-comparable ranking, e.g., where an LER value of 1.5 indicates a intercrop is 50% more efficient that a monocrop. The challenge is to do likewise with each agroecological mandate.

Take the case of stream pollution. Water quality can be scaled by contained pollutants, but the raw data has little outward significance.

Pure water can be used as the zero base, water runoff from a monocrop the comparative point. Assigning monocrop runoff a value of 1.0 resolves at least the conceptional side of the scaling problem.[8]

To avoid an introduced bias in making actual comparisons, one step on the water ranking scale should equate with one step on a biodiversity, esthetic, or some other scale. The problem requires finding what high level of, for example, natural biodiversity expressed in species numbers and populations, would equate with clean water.

There is yet another scaling incongruity, a directional problem. For water runoff, higher values equate with more pollutants, for biodiversity, a higher number means more diversity. Uniformity may require inverting some ranking so they directionally align.

The above demonstrates some of the difficulties encountered. A few mandates conform well to this form of index scaling, others fit less well.

In addressing the issues, it should be mentioned that existing indices, although employing different base points and scale rankings, are helpful. Some of these are economic and achieve cross-comparability through monetary units. This is not advocated here, but the methodologies employed can underwrite scaling determinations, especially when it comes to reconciling diverse mandates.[9] Without delving into all the probable mandates, the selection below illustrates the underlying dilemmas and what might be achieved.

Native Flora and Fauna

For protecting and promoting native flora and fauna, the fluent course is an single index. This is a landscape function or, if a single plot is being evaluated, the assessment is made incorporating the plants and ecosystems in the immediate neighborhood.[10]

The problem is that individuals can have wildly differing views on the ideal, pro-agriculture, pro-wildlife landscape. The difficulty lies in that, when one or a group of natural species gains, others may go into decline.[11] These natural dynamics will bias the outcome unless a species neutral base point is established.

There are indications that territorially formulated indices, one or more per region, are possible. This comes by delving deep into history, finding out what constituted a natural ecosystem during pre-human antiquity. This must answer questions on what early Europe, Asia, the Americas, and other regions once looked like. The answers come with pleasant surprises, e.g., when the distant past seems to have been a highly dynamic and impacted landscape, not something that was static and steadfast across the unpeopled millenniums.[12] Agricultural landscapes are also highly impacted which helps when reaching compromises.

In some regions, adjusting the mix of bats and birds, including the number of species, may involve widening or slightly shifting the range of available ecological niches. This can be done on a number of fronts, ranging from the insertion of more trees to added bat and birdhouses. The base point would be the original unmodified landscape.

Parallel with this is a breakdown into flora and fauna components. The number of supportable native plants is a comparatively easy measure.[13] This will be linked with the fauna evaluations. The agricultural standing of each animal or bird species adds another twist and, with the start of LER analysis, one that should be underway.

For regional birds or animals, one can look to keystone species, if any, or forecast what each individual species can achieve. Both pro- and anti-agricultural species are included. In eastern North America, wrens, bluebirds, and skunks are agriculturally favorable. The sparrow, robin, and the rabbit, all of which are crop detrimental, should be directly discouraged through an adverse habitat or through a landscape that encourages their predators, e.g., foxes that eat rabbits.[14]

Indices that track the progress of individual species can be categorical, i.e., good, fair, or poor surroundings, or quantitative, based on how well the various species benefit or harm crop productivity. The latter is ready made, with a base point (the current levels of crops destroyed or saved by local fauna) and measurable changes in output (as the plots and/ or landscape are modified).

Esthetics

Natural landscapes are a thing of beauty, farm versions can be an equal. In striving for esthetic improvement, plots can be decorated and landscapes can be artistically expressed in both content and placement. Flowing from this is the notion of the edible landscape where cultures, across time, have championed the mixing of productive and decorative plant species.[15] As shown in Photo 15.2, visual relief is achievable within productive constraints.

Esthetics, as a economic concept with an assigned value, is problematic. As the expression goes, beauty is in the eye of the beholder. Accordingly, a base point may be beyond reach. The best one can hope for is to represent this intangible with some highly subjective numeric grading or through an excellent, good, fair, and poor categorization.

Cultural Agroecology

There is another interceding extraneous and intuitive variable, the effect of society and culture on agroecology. It was shown (Chapter 6) that net present value (NPV) analysis is part cultural, less appropriate

PHOTO 15.2. Farm esthetics, a plot seeded with wildflowers and well laid out, picturesque pasture.

when societies have a very long-term view. A greatly understated influence, this translates into initial action through plot content and the physical layouts of farms.

There are many examples of cultural or individual limitations on agroecology. Some are direct, organic gardening may not be profit driven or due to resource constraints, but can be embraced as part of a system of beliefs. Others are less direct, such as when agrotechnologies or variables thereof become subtle expressions of conviction.

Society and cultural values enter the analytical picture early, starting when deciding which agrosystems and farm layouts are to be the subject of attention. Societal inputs continue as intangible, non-quantifiable variables until all has been resolved. In an analytical setting, these will be pre- and post-decision criterion mostly, generally outside the numerical core.

BIOECONOMIC MODELING

Computer models find use in agriculture. The most common applications are in yield prediction or to measure some other outcome.[16] The most prominent, as one might surmise, are models that focus on monocropping.[17]

Yield prediction is only one aspect. If modeling is to offer full economic insight, it must have the ability to do more that predict. It is not a large step beyond prediction for a model to optimize plant spacing, timing of planting, and so on; all leading to the most pressing prognosis, that of profitability. Bioeconomic models, in their most elementary form,

(a) forecast profitability.

In their more advanced forms, these:

(b) optimize profits,
(c) are multi-objective,
(d) are capable of finding multiple solutions.

Profit optimization is done through an iterative search over the respective variable ranges. The yield variables and costs of such determine the profit projections (one or more).

For agroecological purpose, bioeconomic models should be multi-objective, besides optimizing revenue, costs, and hence profit, these should incorporate risk, environmental factors, and other agroecological mandates. This is a far greater step up the complexity scale.

For any complex problem, it is not unusual to have more than one simultaneous solution. For example, the variables may align to produce separate revenue and a cost-oriented profit optimal. A search of the full variable range may pinpoint even more.

The best models will offer multiple solutions, some optimal, others slightly sub-optimal, but well scattered across the design variables. This allows users a worthwhile choice when the non-numeric agroeocological mandates are brought into play. It is not difficult, when doing iterative searches, to program the computer to find the best solution mixes across the variable range.

Trying to gather and position that presented in earlier chapters in a yield and economic prediction model, especially one involving bio-complex intercrops, is a weighty task. This is made more difficult if a model is optimization capable. Although the promise is great, the obstacles are daunting. The full scope, given the number of bicultural systems alone, may prove too big without a paring of the possibilities.

Paring

Given the vast amount of material, it is very difficult, without some paring or reducing factors, to design the best possible agroecosystems and farm landscapes. In curtailing the vast list of possibilities, it is best to limit the options to certain vectors. What might be termed green revolution approach to conventional agriculture relies, to a very large degree, upon the genetic vector coupled with the ex-farm vector. Given the amount of published of research on these two vectors alone, one can extrapolate on the work needed to bring intercropping and agroforestry to similar levels of attainment. From this, the task ahead becomes apparent.

Most agricultural modeling efforts have centered on common crops and common agrosystems. Even with these limiting variables, working monocrop models are not simple efforts, considerable data is needed for development and testing. Generally, the effort is well rewarded. These have proven accurate in normal usage where weather and soil data is available.

As might be expected, mixed species versions can be along species or agrotechnological lines. The first has set species, e.g., maize with sorghum, soybean with oats, cotton with garlic, etc., but within whatever agrotechnologies the mix will support. The second has focuses on a single agrotechnology, e.g., shade, strips, etc., but incorporating a range of crop species and, if appropriate, one or more facilitative plants.

Multi-objective Analysis

A solution is seldom single purpose. Profit may be all-important to accountants, bankers, and absentee landlords, those distant from the site. For those with an intimate association, there is more and compromises are often made. Slightly lower profits are accepted if certain intangibles and/ or agroecological mandates are judged consequential. The profit-environmental tradeoff is one of many such areas. Sustainability and risk reduction can invoke a profit compromise in the preferred outcome.

With a particular agroecosystem, if landusers understood more of what was operationally involved and were certain in an predicted end result, more could be done. This may be the case with subsistence farmers. This is a group that wants assured foodstuffs, now and in the future, then profit, after which come other concerns. Although risk-free food is the primary objective, long-term sustainability, a diversity of foodstuffs, taste, storability, along with religion and culture, all enter the picture.

It is a bit easier to find the solution set for commercial operations where the first concern is profitability, initially and in the future. Further down the list are other considerations. The overriding need for a level of profitability keep the solutions with a narrower range than otherwise the case.

Bioeconomic models layout their solutions in varying forms. Theses can be multi-objective through (1) a single equation or with (2) multiple or tiered equations. The tiered method subdivides into (a) tiebreaking and (b) tradeoffs with limits.

Single Equation

A single equation format requires a prescribed scaling of the objectives, including a quantitative weighting as to their importance. For example, the first object, often profit, may be twice as important as the second. The third may be one-fifth as important as the second. So stated, the computer returns the solution or solutions that are optimal or near optimal to what the objection equation mathematically voices.

A working equation, one utilizing a different weighting of objectives, might be formulated as

$$\text{Maximize } \{0.5\ (\text{profit}) + 0.3\ (\text{TA}) + 0.2\ (1/\text{stream pollution}) + \cdots\}$$

In the above, maximizing the threat assessment (TA) minimizes risk. The pollution scale is also inverted (i.e., less pollution being the small number).

The main obstacle to the single equation format is that some of the agroecological mandates are intangible, lacking a numerically secured comparative index. Given the state of the art and agroecological necessities, the single equation format may be a bit too formal for use outside of well-understood commercial setting (Table 15.1).

TABLE 15.1. Multi-Variable Selection (the Tiebreaker Method)

	Solution sets		
Decision variables	*Agrosystem 1*	*Agrosystem 2*	*Agrosystem 3*
Profitability	$420	$420	$420
Risk (RI)*	**0.9**	**0.9**	1.2
Stream pollution	**0.8**	1.0	1.2

* The RI is so adjusted such that single species monoculture (not among the final choices) has the 1.0 value. Values less than this are less risky.

Tiered Analysis

The multiple or tiered analysis format offer methodological alternatives. One, tiebreaking, employs an ordering where only those solutions that maximizes or near maximizes the first tier criteria are considered. If there is more than one, the second most important criteria serve as a tiebreaker. If there are still multiple choices, a third tier criterion is employed. Agroecological mandates can be second or a third tier criteria, again indexed or not, but most effort will be on quantifiable variables.

Table 15.1 illustrates a decision breakdown of the tiebreaker method. The three simultaneously most profitable solutions are shown (agrosystems 1–3). From these, two offer lesser risk and a third pollutes streams the least. It is this third (agrosystem 1) that is judged the best.

The main obstacle with tiebreaking is that the number of operative tiers is small, meaning fewer decision variables enter the analysis. Again, this format is strongest with commercial farms but, in deferring to second and even third tier criteria, still indulges at least a few environmental worries.

In general, the more complex the farm situation, the greater the likelihood of multiple ties and multiple tiers. This may limit tiebreaking to full farm, rather than plot only analysis.

The second of the multiple equation formats involves tradeoffs with limits. For the first criteria, e.g., profit or assured yields, only solutions considered are those that return a minimum acceptable value. With luck, a lot of solutions will be found. At this point, the second or third tier criterion comes into play. These criteria, often more than three, do not have to be precisely stated. Four or five tiers down, these may only be crude quantitative estimates. As with the upper tiers, the lower tiered criterion must exceed a generously set low bound, otherwise the number of solutions, and the effectiveness of the analysis, will be curtailed.

With a well-formulated series of objectives, the number of solutions should eventually fall below 10. From this, the user makes the final selection without computer guidance; a decision based on which combination of criterion are most farm and landscape appropriate or by appending intangibles (e.g., esthetics, societal and cultural values, etc.) to converge on a unique solution. Being the most flexible and accommodating, this format addresses well the multiple-objective nature of agroecology. The concluding chapter (Table 16.1) offers this in illustration.

Solutions

Although the interaction of many objectives can be expressed as mathematical equations, finding the best single and multi-plot solutions may be more an art than a science. The complexity lies in range of possibilities and insuring that these are well distributed across the sweep of the design variable continuums.

There are other complications. It can be argued that farms are not stable entities. Weather, market prices, as well as soil conditions are in flux. As a result, farmers are chasing, but never reaching, optimal solutions. Also, without ample, trustworthy data, solutions are less a mathematical expression, more an appraisal. It is up to the user to sort through what a bioeconomic model suggests, rejecting most, eventually settling upon the multi-objective solution set deemed most expedient. Given ever-changing events, this represents more a starting, less an end point.

SOLUTION THEORY

Answers to overall agroecosystem questions are more than stochastic end results. In composite, these come together into solution theory.

Going back to the overriding axiom presented in Chapter 1, there is no one-best direction, no one-best way to do things. With no one-best solution, it follows that cropping systems, across individual plots, farms, and regions, each facing different growing and socio-economic conditions, should present a range of optimal solution sets.

If not, and many identical agroecosystems are found, notwithstanding a variation in the soils and topography, under a no-one-size-fits-all hypothesis, many of the plot design formulations are clearly sub-optimal. This provides evidence of agroecological shortcomings and missed opportunities.

Under solution theory, optimal solutions for each farm would be an individual undertaking. Each farm should be noticeably different from neighboring farms when differences in objectives, topography, soils, input constrains, etc., come into play. The stated hypothesis under solution theory is that a more efficient agroecological landscape is expressed in terms of a greater variability in the plot contents and layouts for each contained farm.

ENDNOTES

1. A range of tillage data, including these numbers, can be found in Kaval (2004).
2. Some of the mathematical pitfalls and adverse happenings, those that will prejudice the analytical outcome, are discussed in Vandermeer (1989, pp. 15–28).
3. The most difficult gap to close is a farm that barely provides but, by converting to wood-producing trees, would be financially more remunerative. In this, the worst scenario is where (1) the primary species, often a highly competitive tree (e.g., pine or eucalyptus) discourages many forms of agroforestry and (2) is grown in a 25-year rotation. One solution, a less competitive tree species (e.g., poplar), along income-generating agroforestry, could lessen the gap, both in duration and severity.
4. The stripling planting method can be undertaken with both hardwood and softwood (coniferous) species, the long stems method is more species restricted.
5. This analysis, that of alternate food sources as condition worsen, is often overlooked. The use of such crops goes back into recorded history (e.g., Bainbridge, 1985) and has been documented in Africa, e.g., Harris and Mohammed (2003) list 55 species that Nigerian farmers employ in famine situations.
6. Along these lines, risk analysis is a speculative but, with the available evidence, a safe line of development. The resulting hypothesis is that all rural-dwelling peoples and cultures desire food security and devise means to reduce the threat of famine. The mechanisms are many and varied, lacking only (a) where the situation is on the extreme margin with no opportunity for a backup source (e.g., in desert climes) or (b) where events have engulfed a current situation. Examples of the latter include a change in climate or population growth that has preempted and rendered null the backup source (e.g., the native trees that offered a famine-relief fruit have been cut down for firewood and additional farmland).
7. Cross-farmer cooperation was discussed by Andersson *et al.* (2005), although not in a multi-crop or rotational context. The requirements of such systems are from Watanabe (1992). A hypothetical rendering, showing that compromises can favor both parties, is in Wojtkowski *et al.* (1988).
8. This approach, that of scaling runoff around a single base crop, is found in Finizio *et al.* (2005). The 1.0 base-point scale advocated here is mathematically convenient. As shown, the resulting scale can be inverted without changing the base point. Also, if an inherent scale is not linear, there is some opportunity to fix this. The equation $a^x = b$ where a is a non-linear scaling, b is the linear version, is an expedient method.
9. Monetary units can be comparatively utilized, e.g., Gollin *et al.* (2005). This is not advocated for two reasons: (1) properly formulated units, as with the 1.0 base scale, are easier to compare and (2) by assigning a fiscal worth to quality of life or environmental mandates that are outside the financial sphere, this fosters the notion that these are marketable items. In keeping these separate, tradeoffs are made, but with the goal of finding the best of all worlds. This is quite different from one-for-one tradeoffs, i.e., a revenue dollar for an environmental or quality of life 'dollar.'
10. Addressing the conservation of natural flora and fauna is generally presented as a landscape function, e.g., Swift *et al.* (2004).
11. Carney *et al.* (2005) offer a window into the wildlife base-point issue.
12. There are lot of references to pre-human landscapes. Among those addressing the broader regions are Vera (2000), Mann (2005), and Elvin (2004). Also insightful in this debate are Charles (1997) and Barlow (2001).

13. A biodiversity index was proposed by Coffey (2002).

14. In a useful aside, crop-eating birds can be problematic, especially since there is so little development of controls that do not harm the birds themselves. The open space, a fear of predator mechanism, may not be effective in all regions, e.g., the Northeast US where such birds are aggressive. Other control possibilities include sticky perches (the feet of birds are coated making it hard to land on crops), overhead kites shaped like predators (the sail device, not the type of bird), decoy crops, and bird-resistant crop species (e.g., long awns on grains). The latter is from Harrison (1775).

15. There are numerous reference for esthetic plots and landscapes. To mention a few involving plot expression, there is Beale and Boswell (1991), Wilson (1994), and MacDonald (1994). In a sampling of edible landscape citations, there is Racine *et al.* (1987), Jashemski (1987), McLean (1980), Moynihan (1979), and Husain (2000).

16. A general discussion of ecological modeling, one that equally applies to agroecology, is in Ives (2005). To name a few monocrop and monocropping related models, there are those that predict yield, for maize (e.g., CERES-Maize, CORNF, VT-Maize), for cotton (e.g., GOSSYM) or wheat (CERES-Wheat). There are general yield models (e.g., CROPGRO) and some the look at various effects, e.g., NGAUGE (nitrogen efficiency), SALSA (ecological sustainability in agriculture), STONE and SOILNDB (nitrogen leaching). For intercrops, less has been done. There are approach discussions (e.g., Spitters and Kroff, 1989; Thornton *et al.*, 1990) and a millet–cowpea model (Lowenberg-DeBoer *et al.*, 1990). A limited agroforesty example is by Wojtkowski *et al.* (1991). There are also some economic assessment models for agroforestry use (Graves *et al.*, 2005).

17. Monocropping models do have proven accuracy over a wide range of climatic situations. However, these are not all end-all, there are problems at the fringe, e.g., Carberry and Albrecht (1990) and Mbabaliye and Wojtkowski (1994).

16 Summary

The previously stated axiom, there being no one-best cropping solution, no one-best way to do things, carries the presumption that a number of alternatives or options can be equally profitable. There is no reason that cost orientation should be any less profitable as compared with revenue orientation. Studies on agroecological changeover do support this presumption.

When switching to a more agroecologically based commercial farm practice, some have reported 10–15% less productivity but, when costs are included, overall profits declined only about 4%. Others found production 4% less, with no change in the financial margin. The most striking changeover caused yield reductions of approximately 40%, but with an increase of 40% in the revenue-minus-costs returns.

Another account put the profit margins from agroecology-raised maize at $511 per hectare against $72 for conventional agrochemical-based methods. Similarly, the margins for agroecologically grown soybeans were set at $240 vs. $132 for conventionally propagated soybeans. The comparative figures for alfalfa are $495 vs. $51 in favor of agroecology.[1]

With only a few reporting, a good economic aftermath is far from assured. The problem still remains, finding the combinations of species, spatial patterns, and other design variables to achieve the best results. It is the process of getting to and beyond this point, all while maintaining a positive economic picture, that is of interest to economists.

When studying or applying agroecology, one can start at the beginning, with vectors, or, depending on the known parameters, enter the process at some intermediate step. The intermediate steps are agrobiodynamic principles, implementation concepts, or a process somewhat shortcutted through well-founded agrotechnologies. This chapter explains, in summation, what underwrites the path toward achieving specified economic and non-economic objectives.

AGRICULTURE AND AGROECOLOGY

Chapter 1 presents philosophical views on the relationship between agriculture and agroecology (as in Figure 1.1). One of these, the assertion that agroecology is a subset of agriculture, was discarded.

By default, agroecology is either a parallel expression of agriculture, replete with its own methods and analysis, or it is umbrella discipline under which agriculture resides. For either expression, traditional economic breakdowns do apply, but in restrained measure. Other types or forms of analysis complete the picture.

This re-expression, as disparities between entrenched agriculture and emerging agroecology, occurs throughout this text. From the agroecological perspective, these differences are that[2]:

- a crop species is a component in a larger agroecosystem, one formed by purposeful companion species, undesired plants (i.e., weeds), fauna (wanted or otherwise), and a host of microorganisms;
- each plot is part of, and should interact with, other on farm or regional natural ecosystems and agrosystems;
- there is a longer time frame in planning and implementation;
- there is an awareness, in addressing a broader picture, that agroecology comes replete with some unusual, seldom seen, agricultural practices that address agricultural problems in crop-hostile environments;
- there is a cognizance of native flora and fauna and do-no-harm admonition to these organisms and the natural ecosystems in which they may reside;
- there is an interest in improving the quality of life for people and animals in the immediate neighborhood and in the surrounding region;
- forestry and agroforestry are, or can be, as important to farm landscape as agriculture in terms of monetary returns, the environment, esthetics, and in many other ways;
- there are a wealth of theories, concepts, and principles from which to proceed when data is lacking and experience is limited;
- profit is less a driving force, a lot more, including societal, cultural, and environmental intangibles, are involved.

THE PARADIGM

For conventional farmers, i.e., monocroppers, there are powerful economic forces encouraging the status quo. Banks and other lending institutions, and even governments, have a predilection for the one-crop-per-one-plot scenario.[3] For banks, and the like, the unencumbered ease of yield prediction is viewed as a positive. For governments, the high yields gained through high levels of ex-farm inputs provide more and a greater opportunity to collect taxes. These factors put an indirect, but prevailing pressure, in the revenue-oriented direction.

For the research community, the statistical simplicity of the monoculture is all to enticing. For farmers already faced with very complex operating environment, polycultures and other ecological inputs add another, often analytically unwelcomed, set of design variables.

Despite enticing mono-simplicity, there is much to be gained utilizing what nature offers for free. It is quite possible that some, if not all, of non-natural inputs, i.e., fertilizers, insecticides, etc., can be dispensed with. Although the evidence is scattered and that available comes with no guarantee that proposed practices have been economically optimized, it does suggest that changeovers can be done without overly harming farm profitability.[4]

VECTORS

One of the contrasting points between entrenched agriculture and emerging agroecology is the underlying principles and concepts. In ecology, it is recognized that complex ecosystems cannot be represented by one set of concepts or equations. Instead, these breakdown into series of interrelated mechanisms. How they interact describes the ecosystem.[5] This holds true in agroecology. Vectors and agrotechnologies are manifestations of the mechanistic breakdown.

In addition to understanding, at the most basic and highly abstract starting points, vectors point out gaps or missing opportunities in the vast mosaic of agroecology. In having so many tools from which to proceed, this strongly suggests that there are many routes to productive, profitable, and environmental-friendly agroecology. Without doubt, many have yet to be identified and their utility exploited.

POLICY

The motives of governments do not always coincide with those of farmers. The impetus of the green revolution had less to do with individual farm economics, more to do with providing cheap food and industrial feedstock.[6] Still, if agriculture is to succeed, it should provide constancy and be less risk prone. It is not good if farms and farmers fail (or starve).

Rather than looking across all of agroecology for cropping alternatives, governments offer price supports, direct purchases, crop-loss insurance, land set-asides, and other payments. These policies make conventional agriculture less risky, more appealing, but with a huge and distorting impact on agroecology.[7]

The net effect of policy is often to freeze things as they are. Of course, change is inevitable. The productive successes of a conventional mono-approach does mask sustainability concerns, environmental failures, and some not-so-great economics. As perceptions change, so should policy.

Within these perceptions, there can be grander forces at work. That is what happens when regional agriculture shrinks, i.e., when, through greater per area yields, less land is required to support population levels. The wise course is not to build housing and the like upon prime, but unused farmland. It may be better to keep this in reserve as a nature preserve and a future agricultural assets.

The other option is to de-intensify production, i.e., shift from revenue to cost orientation, utilizing more of the land to produce at a lower per-unit cost.[8] Range farming in the upper mid-west of North America is an example of the latter. Some advocate removing cattle and substituting native bison. This is a switch from cost-oriented grazing to an even more cost-oriented, semi-husbandry form of grazing.

In other regions, the push is in the opposite direction. Greater population pressures are forcing more land into production. The call is for revenue-oriented solutions, often involving a move away from traditional, ecosystem destructive, slash-and-burn techniques. This also might involve keeping agricultural sprawl away from erosion-prone hillsides and other, less suitable lands or, if the lands must be utilized, employing the best possible land-preserving agroecological techniques.[9]

THE AGROTECHNOLOGIES

The agrotechnologies are an encapsulation of what is known. This allows landusers to apply directly, or with modification, tried-and-true techniques. A popular step, many useful agrotechnologies been presented throughout this text.

A listing of recognized agrotechnologies, presented by category, is:

- Productive agrotechnologies
  - Monocultural
    - Pure
    - Varietal/ genus
  - Productive intercropping
    - Simple mixes
    - Strip cropping (seasonal)
    - Barrier or boundary
    - Complex agroecosystems (without trees)

- Productive agroforestry
    - Isolated tree
    - Alley cropping (treerow)
    - Strip cropping (mixed tree)
    - Agroforestry intercropping
    - Shade systems (light)
    - Agroforests
- Facilitative agrotechnologies
  - Perceived monocultures
  - Facilitative intercropping
    - Simple mixes
    - Strip cropping
    - Boundary
    - Covercrops
  - Facilitative agroforestry
    - Parkland
    - Protective barrier
    - Alley cropping (hedgerow)
    - Strip cropping (woody)
    - Crop over tree
    - Physical support systems
    - Shade systems (heavy)
- Rotations
  - Single rotations
    - Series rotations
    - Overlapping cycles
    - Taungyas
      - Simple
      - Extended
      - Multi-stage
      - End stage
    - Continual
- Genetic, varietal, and locational
  - Elevation
  - Scattering
- Land modification
  - Absorption zones/ micro-catchments
  - Infiltration contours
  - Terraces
    - Stone
    - Earthen
    - Progressive
  - Paddies
  - Ponds

- Gabons
- Waterbreaks
- Cajetes
- Water channels
- Mounds and beds
- Stone clusters

Landscape structures

- Windbreaks
- Anti-insect barriers
- Habitats/ corridors
- Riparian buffers
- Firebreaks
- Living fences

Ex-farm

- Irrigation
- Traps

Microbial and environmental setting

- Composting
- Tillage
- Beehives
- Bird and bats

AGROECOLOGICAL ECONOMICS

Agroecological economics is no more than a series of decision tools. With conventional agriculture, persuasion and direction comes through strong data and well-founded analysis. Solutions are often based on a single number.

This will not be the case in agroecology. The unfortunate situation is that, except in scattered instances, there is often not enough information to do the economic sums.

In the short term, not much can be done to improve the informational background. Until this day arrives, decisions must be made along the theoretical, conceptional, intuitive, and experience fronts.

Having said this, practitioners still need information and, although not always in a readily applicable form, enough exists so adequate progress can be made. Outlined below is the suggested economic pattern of agroecology. Keeping in mind that each practitioner has their own situation and agenda, these steps come laden with caveats. Although the subject to twists and turns, exceptions and acceptors, certainties and unknowns, there is, underlying all, a systematic progression.

The Agrotechnologies and the Outcomes

In deciding which agroecology or agrotechnologies are best, the end-use scenarios come into play. If hillside erosion is a problem, good erosion control would be high on the list. This is often done through the desirable agroecosystem properties (DAPs). Before compiling a detailed DAP listing, one might first weight the pros and cons of the standard-form, principal-mode agrotechnologies (see the Appendix). The agrotechnological descriptions (Chapter 4 through 11) facilitate the elimination process.

Along the way, it must be realized that, in their standard versions, the agrotechnologies are seldom site and user optimal. Design modifications, often species, spacing, spatial pattern, and timing, can make these more site and user suitable. If still not entirely befitting, a range of auxiliary and add-on agrotechnologies can change the standard-form DAPs. This makes the process more complex while, at the same time, offering agrotechnologies that address all or most concerns.

Resource-Use Efficiency

Many of the productive agrotechnologies have undergone enough development so that a preliminary economic distillation can be done, answering the yes-or-no question on whether it has profit potential for a given use situation. For example, one would generally choose a highly revenue-oriented agrosystem for a near-city, high-input, market garden.

Part of this picture is essential resource-use efficiency, how well are the light, water, and nutrients being utilized by the system. More consequential, how well are the limiting site resources being utilized. For this, the preferred analytical tool is the land equivalent ratio (LER). To be complete, product selling prices should be included (through the relative value total, RVT) as well as labor inputs (by way of the $CER_{(RVT)}$ (cost equivalent ratio)).

Sustainability, the ability of the plot to yield over time, might be part of the analysis. A user may wish to choose a single agrotechnology, one nutritionally supported through ex-farm inputs, achieve sustainability through a rotational sequence, or select a internally, self-sustaining design. It is also possible to utilize some combination of nutrient strategies.

Profitability and Risk

Once optimal or near optimal agrotechnologies are identified, then the potential profitability of each, through financial accounting, is measured. This is not the end all, but provides what may be the most imperative number.

Loosely defined, per area profitability will depend a lot on the economic orientation and inherent in this is the land, labor, and monetary situation of a farmer. If labor and/ or inputs are relatively inexpensive, revenue-oriented agrotechnologies, those that employ less land and more workers or inputs, may be the best course. If not, a cost-oriented path, one using less labor, less inputs, and more land, may be best.

Next to profit, risk may be the second most important decision variable. This is less a factor when governmental programs, e.g., crop insurance, lessens the negative impact of crop failure, critical when farmers face cropping threats without outside help.

Ex-profit Variables

There are a host of tangibles and intangibles that are part of the decision process. Clean water, native flora and fauna, visually pleasing farm landscapes, an agreeable area climate, and the like are part of the complete agroecological picture. Cultural, societal, and religious values are yes-or-no variables. Lacking in the graduated subtilizes of the ex-profit, ex-risk variables, these are inclusive in the agroecological pre-screening.

Multi-criterion Decisions

Following from the axiom; no one-best direction, no one-best way to do things, the ideal, one to strive for, lies in having multiple solutions, all profit certified, from which to choose. Once these are determined to be near or at optimal form, the process of selection turns multi-criterion.

Part of this is an analysis using desirable agroecosystem properties (DAPs). These refine the relationship between site and user needs and economic orientation. These also decide on the degree of inherent risk. Additionally, analysis may examine how well each interacts with nature, which harms less native plants, the fauna of a region, and/ or downstream aquatic ecosystems. With the assurance that each choice is financially profitable, there can be any number of tiebreaker questions.[10]

To illustrate, Table 16.1 offers a decision set with three resource efficient, profitable, and pre-optimized potential solutions. Each of these is evaluated using five after-profit determination variables. Per area profit, as an actual figure, anchors the decision process. Assuming that the problems in formulating cross-comparable indices have been ironed out, i.e., all agree on scale and direction, these indices rank the lower-tier decision criteria.[11]

TABLE 16.1. Multi-variable Selection (the tradeoffs within limits method)

Decision variables	*Solution sets*		
	Agrosystem 1	*Agrosystem 2*	*Agrosystem 3*
Profitability (set minimum $400)	$420	$450	$400
Risk (RI)	0.85	1.0	0.9
Sustainability	1.5	1.0	1.3
Cropping flexibility	0.8	1.0	1.2
Stream non-polluting	1.8	1.0	1.1
Encourages songbirds	Good	Good	Poor
Esthetic	Good	Fair	Good

The tLER a measure of sustainability.

Solutions

The final decision in Table 16.1 (the choice between agrosystems 1, 2, or 3) is subjective. The hope is that the decision maker is very familiar with the situation and that more than a casual peruse of the data underpins the final choice. The tradeoff are not always one for one and, in the best-case scenario, one or two alternatives will stand out, i.e., a few solution will give high marks for most or all of the decision variables.[12]

If any otherwise favorable agroecosystem has one glaring weakness, this might be rectified through an auxiliary system or by modifying other farm plot. If such an ex-plot remedy is not at hand, another agrosystem would be selected.

In Table 16.1, solution 1 (i.e., agrosystem 1) is weak with regards to cropping flexibility, i.e., an ability to change the output mix in response to fluctuating market prices. The farmer could go to the second best solution or accept the flexibility shortcoming, making this variable a higher priority when reformulating some other farm plot.

Steps Beyond

Decision arrays can be the product of comprehensive bioeconomic model or assembled as parts, each developed from a separate analysis. Depending on whether the landscape contains independent or cross-linked plots, the process may continue on into the landscape level.

Closer to the ideal is to treat the landscape as in Table 16.1 but, instead of agrosystems 1, 2, and so on, the landuser chooses from a range of landscape options (farm landscapes 1, 2, 3, etc.) presented in the same optimized, multi-objective, multi-solution format. Still closer to the ultimate ideal is to work across farm boundaries where the regional landscape, one involving many farms, is evaluated and optimized in its entirety (as in regional landscape options 1, 2, 3, etc.).

In reaching the stated goal, one does not have to go the full distance, starting with essential resource efficiency in the agrotechnologies (one or more) first selected, ending at a decision matrix. In all likelihood, many will abbreviate the process. This involves employing, for targeted purpose, analytical elements and analytical segments. In the maize with herbicide-killed weeds case study (Photo 12.1), a good solution was close at hand, the overall problem well contained. In this case, simple substitution and a fairly basic $CER_{(LER)}$ or a cost–benefit analysis would complete the process.

Proposing solutions that change or greatly modify individual agrotechnologies, as in a major redesign, transcends a simple substitution. This should trigger a deeper agro-analysis, one traversing much of, or the full, analytical distance. Warnings come when a complete agroecosystem changeover is contemplated and the process overly truncated. Analytical weaknesses can include not seeking optimization or in not looking far enough afield at some atypical, but conceivably highly site suitable and user salutary alternatives.[13]

CONCLUSION

Agroecological economics is the sum of many parts. Essential resources, use objectives, as well as insects, birds, and a host of other components must come together to form a complete, and encouraging, agrobionomic picture. From this, directions and ultimately cropping options are derived.

Results and solutions can be arrived through many different vector routes. The initial economic challenge lies in sorting the basic plant–plant possibilities, finding the best, and building upon these in the quest for a broad, economically optimal solution(s). The journey often starts byway of yields and through efficiency ratios.

Once some initial solutions are located, more variables, including the DAPs, environmental contributions, native flora and fauna, as well as esthetics, comfort, satisfaction, and host of site and user intangibles, are added. The purpose is to pair what should be a large list of agroecological and agrotechnological possibilities. Traditional economic measures, i.e., revenue and costs, round out and confirm the findings from the decision process.

Given the state of the art, and of the science, selecting and refining the agroecological possibilities are not easy roads to travel. Despite daunting options and the expansive alternatives, the process should not be curtailed. In the search for productive, cost effective, and environmental-friendly solutions, the best results are obtained by looking at and analyzing a full range of possibilities. In this way, it becomes possible to uphold the mandates and the covenants of agroecology.

ENDNOTES

1. This brief foray into the economics of agroecological changeover comes from, in order of presentation, Jordan *et al.* (1996), Jordan *et al.* (1997a, b), IACPA (1998), Mäder *et al.* (2002), and Charles and Clover (1993, p. 120). The comparative profits, from North America; are reported in Sayre (2005). These look at large scale farming. Similar results have been reported for small farms, see Gomes de Almeida and Fernandes (2006) for a Brazilian example. Without a slew of broad, well-directed studies, definitive statements on the economic outcome of an agroecological makeover are not possible.
2. This list is a highly modified restatement, first compiled from Hansen (1996) and Rigby and Cáceres (2001), later from Wojtkowski (2006a).
3. The preference for banks and other lending institutes for monocropping is in Godoy and Bennett (1991) and McNeely (1993).
4. As to profitability, those listed in Endnote 1 provide an unqualified endorsement, i.e., that agroecology can openly compete with conventional agriculture; others, e.g., Jordan (2004) or Brownlow *et al.* (2005), find profitability, but only through the price premium paid to organic producers. In a reversal, some, e.g., van der Vossen, find profitability through versions of agroecology, but without the added costs of organic labeling compliance.
5. The breaking down of complex systems into simpler and underlying mechanisms is discussed by Grimm *et al.* (2005).
6. The motives of the green revolution are in Polak (2005).
7. Policy, as a formidable barrier to agroecology, is mentioned in Dalgaard *et al.* (2003). Some applied policy studies are by Feehan *et al.* (2005), Carney *et al.* (2005), and Wilson *et al.* (2003).
8. Predicting a trend toward cost orientation is Gollin *et al.* (2005).
9. Sampling of agricultural intensification studies are in Keil *et al.* (2005) and Versteeg *et al.* (1998).
10. The prevailing thought is that better conclusions are reached if each option is distilled to no more than five after-profit variables. Normally about seven, instead of the three options as presented here, is considered an optimum number. The use of three options is only a presentational convenience. Steuer (1986) provides a window into multi-criteria decision-making, a field normally applied to business decisions.
11. As stated, the values for below-profit criteria are hypothetical, this avoids the dilemma in having to scale, directionally aline, and insure that units cross-compare. More work is needed on scaled variables.
12. Versteeg *et al.* (1998) example a field problems where multiple solutions are presented.
13. There are drawbacks in going to far afield. Although atypical solutions can be, at times, more profitable and/ or more environmentally suitable, landusers are often reluctant to embrace agroecosystems outside their immediate experience range.

Appendix: A Brief Overview of the Principal-Mode Agrotechnologies

The summaries below constitute comparisons, against the base system, i.e., of a pure monoculture, or against a comparative norm. All descriptions are predicated around the starting or standard design for each principal-mode agrotechnology.

Monocultural – pure

Standard form: Large or small plots composed of a single, genetically alike or similar crop species.

Use schema: The simplest of agroecosystems with wide use possibilities.

Pros

High productivity from a single crop species

Crop flexible

Comparatively easy to manage.

Cons

Expensive in terms of labor and input requirements

Naked to all or most cropping threats

Can be environmentally troublesome if ex-farm inputs productively support and protect.

Monocultural – varietal/ genus

Standard form: Large or small plots composed of one crop type, but with more than one variety.

Use schema: Same as with the pure monoculture.

Pros

Crop flexible.

Cons

Less naked, but still exposed to a wide range cropping threats

Marketing or use concerns when the varietal outputs are distinguishable.

Productive intercropping – simple mixes

Standard form: The interplanting of different types of short-duration (often annual) crops.

Use schema: Best where high-output levels from a mix of species are required and where crops are hand planted and hand harvested.

Pros

High-output levels (as measured through the land equivalent ratio (LER))

Cropping flexible when the component crops are mutually complementary

High to moderate level of threat exposure.

Cons

Requires manual labor

Not always sustainable unless provisions (e.g., rotations) are included in the design.

Productive intercropping – strip cropping

Standard form: Strips of one crop type or intercrop alternating with another crop or intercrop mix.

Use schema: Best where high-output levels are desired and the crops are machine planted and harvested.

Pros

High-output levels

Allows for agricultural mechanization

Protects against erosion and suitability for not-too-steep hillsides

Low to moderate levels of protection against crop-eating insects and plant diseases

Does not require that inter-strip crops (one or more) be mutually complementary, especially if a buffer species is employed.

Cons

For mechanized versions, the cropping area must be strip suitable (allowing extended, unbroken strips).

Productive intercropping – barrier or boundary

Standard form: A comparatively small monocrop or intercrop plot is surrounded by another, short-duration, productive species.

Use schema: Used to ecologically isolate and protect small, hand-maintained plots.

Pros

Cropping flexible

Moderate protection against one or more identified threats.

Cons

Limited to smaller-sized plots.

Productive intercropping – complex agroecosystems

Standard form: A large number of short-duration, productive species are intermixed.

Use schema: Used where multiple, low-input yields are produced with hand labor.

Pros

Potentially high-output levels as measured by the LER

Cropping flexible

Moderate protection against a range of in-season threats.

Cons

Requires hand labor

Multiple outputs limits use (mainly to non-commercial situations).

Facilitative intercropping – simple mixes

Standard form: One or more short-duration or non-woody, non-productive perennials are interplanted with one or more short-duration crops.

Use schema: Employed when the soil nutrient content is poor and must be augmented and/ or to protect against a specific cropping threat.

Pros

Yields can be increased with the correct facilitative pairings

Crop flexible

Moderate to high sustainability

Protection against erosion and other cropping threats.

Cons

Higher costs

Unanticipated interspecies competition can decrease yields

Facilitative plants must be matched against the crop nutrient needs and/ or to counter a specific threat.

Facilitative intercropping – boundary

Standard form: A comparatively small monocrop or intercrop plot is surrounded by another, short-duration, non-productive species.

Use schema: Used to ecologically isolate and protect small, hand-maintained plots.

Pros

Cropping flexible

Moderate protection against one or more identified threats.

Cons

Limited to smaller-sized plots.

Facilitative intercropping – covercrops

Standard form: Placing a ground-hugging, non-woody, facilitative plant beneath one crop or some intercrop.

Use schema: Simple means to ecological augment a monocrop.

Pros

Possible increases in yields

Reductions in weeding costs

Moderate to high protection against specific threats.

Cons

Unanswered questions on the cover species, the timing, and the duration.

Productive agroforestry – isolated tree

Standard form: A single, large, productive tree over a large cropping or pasture area.

Use schema: Most common as shade tree for pasture grazed animals or for the minor ecological services provided to crops.

Pros

Low levels of protection against a few threats

Increased weight gains in the shade provided to cattle and other animals.

Cons

The trees can be an unreturned cost.

Productive agroforestry – alley cropping (treerow)

Standard form: Strips of crops alternating with single rows of a productive tree species.

Use schema: High-intensity productivity on good, level sites.

Pros

High levels of productivity

Low to moderate protection against insects and plant diseases

Ease of harvest for the tree component

Good erosion control properties.

Cons

Cropping inflexibility
Often unsustainable requiring ex-farm inputs
Rows must be oriented to maximize light interception.

Productive agroforestry – strip cropping

Standard form: Strips of productive trees alternating with strips of crops.

Use schema: High-intensity productivity with some inherent agroecological gains.

Pros

High levels of output
Moderate to high levels of protection against all cropping threats
Harvests, especially for high-volume treecrops is expedited.

Cons

With high levels of outputs, there are sustainability issues
Rows must be oriented to maximize light interception.

Productive agroforestry – agroforestry intercropping

Standard form: A mix of productive, often fruiting, trees in an orchard format.

Use schema: Utilized much the same as orchards, but where a mix of outputs is desired.

Pros

Produces over a longer period (good in certain markets and/ or for foraging animals)
Spreads labor (harvest) inputs over a longer time span
Moderate protection against insect or disease threats.

Cons

The market must accept smaller amounts of varying fruit and/ or nut varieties.

Productive agroforestry – shade systems (light)

Standard form: Crops raised beneath a continuous, productive, highly light-permeable tree canopy or beneath a mix of productive tree species.

Use schema: Used in situations where high levels of mixed outputs, mostly from the trees, are desired.

Pros

High-output levels as measured by the LER
Moderate to strong protection against a range of threats
Constant micro-climate
Good erosion control properties.

Cons

Long-term sustainability may be lacking if all the fruits are harvested.

Productive agroforestry – agroforests

Standard form: A jumble of tree, shrubs, and non-woody species most of which provide some useful output.

Use schema: Found where users desire many different outputs are produced at a very low cost over a long time span.

Pros

Highly sustainable

Free from most cropping threats

A very cost effective means of production

Considered the most environmental-friendly agroecosystem.

Cons

The user must be content with low levels of many types of outputs.

Facilitative agroforestry – parkland

Standard form: A few scatter trees over a pasture or cropping area.

Use schema: Provides shade for pastured animals or minor ecological services for crops.

Pros

Allows for the use of farm machinery

Minor agroecological gains.

Cons

Requires one or two role-established tree species

In cropping situations, tree maintenance can be an unjustified cost.

Facilitative agroforestry – protective barrier

Standard form: Trees or shrubs surrounding crop plots or pastures.

Use schema: Encountered as wind barrier with resulting productivity gains or as live fencing.

Pros

The blocking of the prevailing winds most often increases per area output

Moderate protective gains against harmful insects and plant diseases.

Cons

The associated costs must be use justified.

Facilitative agroforestry – alley cropping (hedgerow)

Standard form: Strips of crops alternating with single or double rows of pruned hedges.

Use schema: Used to increase cropping sustainability when the hedge prunings nutritionally assist the crop and/ or to permit hillside cultivation without erosion.

Pros

Permits continued cropping where not otherwise possible (on poor soils or hillsides)

Low to moderate protection against specific cropping threats.

Cons

Labor intensive with pruning during the plowing and planting season

Lacks cropping flexibility as each hedge species is specific to a crop type.

Facilitative agroforestry – strip cropping

Standard form: Crop strips alternating with strips of short-pruned trees, the intent being that the cut biomass from the tree strips fertilizes the crop.

Use schema: Insures high crop yields where soils are poor and farm machinery is employed.

Pros

Protects against erosion with suitability for not-too-steep hillsides

Long-term sustainability as this eliminates the need to supply outside nutrients

Moderate to high protection against a range of cropping threats.

Cons

Land heightened as about one-half the area is taken by one or more non-productive woody species

Requires a tree species with nutrient-rich foliage and coppice survivability

The crop or crops must seek the same balance of nutrients as provided by the tree coppice

Requires continued maintenance so that the trees do not grow large.

Facilitative agroforestry – crop over tree

Standard form: Shrubs or trees, pruned to ground level, located beneath taller crops.

Use schema: Reduces labor and inputs costs in the ecological services provided by the trees.

Pros

Permits light demanding crops to be raised with trees

A cost effective means to provide low to moderate levels of protection.

Cons

The opportunities for good crop/ tree matches are limited.

Facilitative agroforestry – physical support systems

Standard form: Vines are supported below, within, or above the canopy of a productively inert tree.

Use schema: Employed as an alternative to trellises where accompanying ecological gains are desired.

Pros

Facilitative services, in the form of favorable nutrient dynamics, can increase yields

Moderate protection against a range of threats.

Cons

Pruning needs can impinge other farm activities.

Facilitative agroforestry – shade systems (heavy)

Standard form: Crops are raised beneath a unbroken, density, tree canopy.

Use schema: A very low-cost means to raise shade-tolerant crops.

Pros

Weeding costs are greatly reduced

A more or less constant micro-climate is assured

Effective protection against a range of threats

Strong ecological spillover possibilities

Wildlife favorable.

Cons

Best with low-intensity agriculture

Only possible with perennial, shade-resistant crops.

References

Abate, T., van Huis, A. and Ampofo, J.K.O. (2000) Pest management strategies in traditional agriculture: an African perspective. *Annual Review of Entomology* **45**(1): 631–660.

Akyeampong, E., Hitimana, L., Franzel, S. and Munyemana, P.C. (1995) The agronomic and economic performance of banana, bean and tree intercropping in the highlands of Burundi: an intern assessment. *Agroforestry Systems* **31**: 199–210.

Ali, A.M.S. (2005) Homegardens in smallholder farming systems: examples from Bangladesh. *Human Ecology* **33**(2): 245–270.

Alomar, O., Goula, M. and Albajes, R. (2002) Colonization of tomato fields by predatory mirid bugs (Hemiptera: *Heteroptera*) in northern Spain. *Agriculture Ecosystems and Environment* **89**: 105–115.

Altieri, M.A. and Nicholls, C.I. (2004) *Biodiversity and Pest Management in Agroecosystems*, 2nd edition. Food Products Press, New York, 236 pp.

Altieri, M.A. and Trujillo, J. (1987) The agroecology of corn production in Tlaxcala, Mexico. *Human Ecology* **15**(2): 189–220.

Anderson, R.L. (2003) An ecological approach to strengthen weed management in the semiarid Great Plains. *Advances in Agronomy* **80**: 33–62.

Anderson, R.L. (2005) A multi-tactic approach to manage weed population dynamics in crop rotations. *Agronomy Journal* **97**(6): 1579–1583.

Andersson, H., Larsén, K., Andersson, C., Blad, F., Samuelsson, J. and Skargren, P. (2005) Farm cooperation to improve sustainability. *Ambio* **34**(4): 383–387.

Anon. (1872) *Gardeners' Journal*. United Society [Shakers], New Lebanon, NY, 180 pp.

Anon. (2002) Bee off with you. *The Economist* **365**(8297): 78.

Anon. (2003) Alley cropping in Guam; hedgerows benefit crops and soil nitrogen. *Inside Agroforestry,* USDA (Summer/Fall): 7–8.

Ashley, M.D. (1986) *Agroforestry in Haiti*. University of Maine, 69 pp.

Atlin, G.N. and Frey, K.J. (1989) Breeding crop varieties for low input agriculture. *American Journal of Alternative Agriculture* **4**(2): 53–58.

Aumeeruddy, Y. and Sansonnens, B. (1994) Shifting from simple to complex agroforestry systems: an example for buffer zone management from Kerinci (Sumatra, Indonesia). *Agroforestry Systems* **28**: 113–141.

Ayuk, E.T. (1997) Adoption of agroforestry technology: the case of live hedges in the central plateau of Burkina Faso. *Agricultural Systems* **54**: 189–206.

Babu, K.S., Jose, D. and Gokulapalan, C. (1992) Species diversity in a Kerala home garden. *Agroforestry Today* **4**(3): 15.

Bäckman, J.-P.C. and Tianen, J. (2002) Habitat quality of field margins in a Finnish farmland area for bumblebees (Hymenoptera: *Bombus* and *Psithyrus*). *Agriculture Ecosystems and Environment* **89**: 53–68.

Bainbridge, D.A. (1985) The rise of agriculture: a new perspective. *Ambio* **14**(2): 148–151.

Banda, A.Z., Meghembe, J.A., Ngugi, D.N. and Chome, V.A. (1994) Effect of intercropping maize and closely spaced *Leucaena* hedgerows on soil conservation and maize yield on a steep slope at Ntcheu, Malawi. *Agroforestry Systems* **27**: 12–22.

Barlow, C. (2001) Ghost stories from the ice age. *Natural History* **110**(7): 62–67.

Beale, G. and Boswell, M.R. (1991) *The Earth Shall Blossom*. The Countryman Press, Woodstock, VT, 263 pp.

Beer, J. (1987) Advantages, disadvantages and desirable characteristics of shade trees for coffee, cacao and tea. *Agroforestry Systems* **5**: 3–13.

Beets, W.C. (1982) *Multiple Cropping and Tropical Farming Systems*. Westview Press, Boulder, CO, 156 pp.

Belcher, B., Michon, G., Angelson, A., Pérez, M.R. and Asbjornsen, H. (2005) The socio-economic conditions determining the development, persistence, and decline of forest garden systems. *Economic Botany* **59**(3): 245–253.

Benin, S., Smale, M., Pender, J., Gebremedhin, B. and Ehui, S. (2004) The economic determinants of cereal crop diversity on farms in the Ethiopian highlands. *Agricultural Economics* **31**(2–3): 197–208.

Berger, C. (2006) What are bugs worth? *National Wildlife* **44**(October – 6): 36–43.

Bettiol, W. (1999) Effectiveness of cow's milk against zucchini squash powdery mildew (*Sphaerotheca fuliginea*) in greenhouse conditions. *Crop Protection* **18**(8): 489–492.

Beyerlee, D. and Murgai, R. (2001) Sense and sustainability revisited: the limits of total factor productivity measures of sustainable agricultural systems. *Agricultural Economics* **26**(3): 227–236.

Biazzo, J. and Masivnas, J.B. (2000) The use of living mulches for weed management in hot pepper and okra. *Journal of Sustainable Agriculture* **16**(1): 59–79.

Blanford, H.R. (1925) Regeneration with the assistance of taungya in Burma. *The Indian Forest Records* **11**(3): 1–39.

Bradfield, S. (1986) Sociocultural factor in multiple cropping, pp. 267–284. In: Francis, C.A. (ed.), *Multiple Cropping Systems*. MacMillan Publishing Co., New York, 383 pp.

Brenner, A.J. (1996) Microclimatic modifications in agroforestry, pp. 159–188. In: Ong, C.K. and Huxley, P. (eds.), *Tree–Crop Interactions: A Physiological Approach*. CAB International, Wallington, UK, 386 pp.

Brookfield, H. (2002) Agrodiversity and agrobiodiversity, pp. 9–14. In: Brookfield, H., Padoch, C., Parsons, H. and Stocking, M. (eds.), *Cultivating Biodiversity*. ITDG Publishing, London, UK, 291 pp.

Brown, B.J. and Marten, G.G. (1986) The ecology of traditional pest management in Southeast Asia, pp. 241–272. In: Marten, G.G. (ed.), *Traditional Agriculture in Southeast Asia: A Human Ecology Perspective*. Westview Press, Boulder, CO, 358 pp.

Brown, M.W. and Tworkoski, T. (2004) Pest management benefits of compost mulch in apple orchards. *Agriculture, Ecosystems and Environment* **103**(3): 465–472.

Browne, P.J. (1832) *The Silva Americana*. William Hyde and Co., Boston, MA, 407 pp.

Brownlow, M.J.C., Dorward, P.T. and Carruthers, S.P. (2005) Integrating natural woodlands with pig production in the United Kingdom: an investigation of potential performance and interactions. *Agroforestry Systems* **65**(3): 251–263.

Bryan, J.A. (1999) Nitrogen fixation of leguminous trees in traditional and modern agroforestry systems, pp. 161–182. In: Ashton, M.S. and Montagnini, F. (eds.), *The Silvicultural Basis for Agroforestry Systems*. CRC Press, New York, 278 pp.

Budelman, A. (1988) The performance of leaf mulches of *Leucaena leucoephala*, *Flemingia macrophylla* and *Gliricidia septium* in weed control. *Agroforestry Systems* **6**: 137–146.

Budelman, A. (1990) Woody legumes as live support in yam cultivation: the yam – *Gliricidia sepium* association. *Agroforestry Systems* **10**: 61–69.

Caborn, J.M. (1965) *Shelterbelts and Windbreaks*. Faber and Faber, London, UK, 287 pp.

Cálix de Dios, H. and Castillo Martinez, R. (2000) Soportes vivos para pithaya (*Hylocereus* spp.) en sistemas agroforestales. *Agroforesteria en las Américas* **7**(28).

Caramori, P.H., Androcioli Filho, A. and Leal, A.C. (1996) Coffee shade with *Minosa scabrella* Benth. for frost protection in southern Brazil. *Agroforestry Systems* **33**: 205–214.

Carberry, P.S. and Albrecht, D.G. (1990) Tailoring crop models to the semiarid tropics. In: Muchow, R.C. and Bellany, J.A. (eds.), *Climatic Risk in Crop Production: Models and Management for the Semiarid Tropics.* CAB International, Wallingford, UK, pp. 157–182.

Carney, P.D., Manchester, S.J. and Firbank, L.G. (2005) Performances of two agri-environmental schemes in England: a comparison of ecological and multi-disciplinary evaluations. *Agriculture, Ecosystems and Environment* **108**(3): 178–188.

Carter, J. (1996) Alley farming: have resource-poor farmers benefited? *Agroforestry Today* **8**(2): 5–7.

Castelán-Ortega, O., González, E.C., Arriaga, J.C. and Chávez, M.C. (2003) Mexico, pp. 249–269. In: Brookfield, H., Parsons, H. and Brookfield, M. (eds.), *Agrodiversity: Learning from Farmers Across the World.* United Nations University Press, Tokyo, 343 pp.

CEPIA (1986) *Technologias Campesinas de los Andes del Sur.* Centro el Canelo de Nos, Perú, 278 pp.

Charles, D. (1997) Can bison replace engines of war? *New Scientist* **154**(10 May – 2081): 10.

Charles, D. (2002) Fields of dreams. *New Scientist* **173**(5 January – 2324): 25–27.

Charles (Prince of Wales) and Clover, C. (1993) *Highgrove: An Experiment in Organic Gardening and Farming.* Simon and Schuster, New York, 288 pp.

Chaucer, G. (1382) *Parliament of Foules.*

Chesney, P., Schlönvoigt, A. and Kass, D. (2000) Producción de tomate con soportes vivos en Turrialba, Costa Rica. *Agroforesteria en las Américas* **7**(26).

Chifflot, V., Bertons, G., Cabanettes, A. and Gavaland, A. (2006) Beneficial effects of intercropping on growth and nitrogen status of young wild cherry and hybrid walnut trees. *Agroforestry Systems* **66**(1): 13–21.

Chou, C.-H. (1992) Allelopathy in relation to agricultural productivity in Taiwan: problems and prospects, pp. 179–203. In: Rizvi, J.H. and Rizvi, V. (eds.), *Allelopathy: Basic and Applied Aspects*. Chapman and Hall, London, 490 pp.

Cleveland, C.C., Townsend, A.R., Schmidt, S.K. and Constance, B.C. (2003) Soil microbial dynamics and biogeochemistry in tropical forests and pastures of southwestern Costa Rica. *Ecological Applications* **13**(2): 314–326.

Coffey, K. (2002) Quantitative methods for the analysis of agrobiodiversity, pp. 78–96. In: Brookfield, H., Padoch, C., Parsons, H. and Stocking, M. (eds.), *Cultivating Biodiversity*. ITDG Publishing, London, UK, 291 pp.

Coghlan, A. (2006) Catch every drop to fight world hunger. *New Scientist* **19**(26 August – 2566): 14.

Cooper, P.J.M., Leakey, R.R.B., Rao, M.R. and Reynolds, L. (1996) Agroforestry and the mitigation of land degradation in the humid and sub-humid tropics of Africa. *Experimental Agriculture* **32**: 235–290.

Dagar, J.C., Singh, N.T. and Singh, G. (1995) Agroforestry options for degraded and problematic soils of India, pp. 96–120. In: Singh, P., Pathak, P.S. and Roy, M.M. (eds.), *Agroforestry Systems for Sustainable Land Use*. Science Publishers, Lebanon, NH, 283 pp.

Dai, Q., Fletcher, J.J. and Lee, J.G. (1993) Incorporating stochastic variables in crop response models: implications for fertilizer decisions. *American Journal of Agricultural Economics* **75**: 377–386.

Dalgaard, T., Hutchings, N.J. and Porter, J.R. (2003) Agroecology, scaling and interdisciplinarity. *Agriculture, Ecosystems and Environment* **100**(1): 39–51.

Damschen, E.I., Haddad, N.M., Orrock, J.L., Tewsbury, J.J. and Levey, D.J. (2006) Corridors increase plant species richness at large scales. *Science* **313**(9 September – 5791): 1284–1286.

Dapaab, H.K., Asafu-Agyei, J.N., Ennin, S.A. and Yamoah, C. (2003) Yield stability of cassava, maize, soya bean and cowpea intercrop. *Journal of Agricultural Science* **140**: 73–82.

Davis, J.H.C. and Garcia, S. (1983) Competitive ability and growth habit of indeterminate beans and maize for intercropping. *Field Crops Research* **6**: 59–75.

Davis, J.H.C., Roman, A. and Garcia, S. (1987) The effects of plant arrangement and density in intercropped beans (*Phaseolus vulgaris*) and maize. II. Comparison of relay intercropping and simultaneous planting. *Field Crops Research* **16**: 117–126.

De Costa, W.A.J.M. and Surenthran, P. (2005) Tree–crop interactions in hedgerow intercropping with different tree species and tea in Sri Lanka. 1. Production and resource competition. *Agroforestry Systems* **63**(3): 199–209.

de Foresta, F., Basri, A. and Wiyono (1994) A very intimate agroforestry association cassava and improved homegardens: the Mukibat technique. *Agroforestry Today* **6**(1): 12–14.

de Wit, C.T. (1992) Resource use efficiency in agriculture. *Agricultural Systems* **40**: 125–151.

Dedek, W., Pape, J., Grimmer, F. and Körner, H.-J. (1989) Integrated pest control in forest management – combined use of pheromone and insecticide for attracting and killing the bark beetle *Ips typographus*. II. Effects of methanidophos treatment following bark penetration into the ascending sap of pheromone-baited spruce. *Forest Ecology and Management* **26**: 63–76.

den Biggelaar, C. (1996) *Farmer Experimentation and Innovation: A Case Study of Knowledge Generation Process in Agroforestry Systems in Rwanda.* Community Forestry Case Study Series 12, FAO, Rome, 123 pp.

Devlaeminck, R., Bossuyt, B. and Hermy, M. (2005) Seed dispersal from forest into adjacent cropland. *Agriculture, Ecosystems and Environment* **107**(1): 57–64.

Dix, M.E., Hill, D.B., Buck, L.E. and Rietveld, W.J. (1997) Forest farming: an agroforestry practice. *Agroforestry Notes* **7**(November): 1–4.

Donald, C.M. (1963) Competition among crop and pasture plants, pp. 1–118. In: Norman, A.G. (ed.), *Advances in Agronomy*, Vol. 15. Academic Press, New York, 415 pp.

Eilittä, M., Sollenberger, L.E., Littell, R.C. and Harrington, L.W. (2003) On-farm experiments with maize–mucuna systems in the Los Tuxtlas region of Veracruz, Southern Mexico. II. Mucuna variety evaluation and subsequent yield. *Experimental Agriculture* **39**(1): 5–17.

Ellis-Jones, J., Schulz, S., Douthwaite, B., Hussaini, M.A., Oyewole, B.D., Olanrewaju, A.S. and White, R. (2004) An assessment of integrated *Striga hermonihica* control and early adoption by farmers in Northern Nigeria. *Experimental Agriculture* **40**(3): 353–368.

Elvin, M. (2004) *The Retreat of the Elephants: An Environmental History of China*. Yale University Press, New Haven, CT, 564 pp.

Emerman, S.H. and Dawson, T.E. (1996) Hydraulic lift its influence on the water content of the rhizophere: an example from sugar maple, *Acer saccharum*. *Oecolgia* **108**: 273–278.

Esekhade, T.U., Orimoloye, J.R., Ugwa, I.K. and Idoko, S.O. (2003) Potential of multiple cropping systems in young rubber plantations. *Journal of Sustainable Agriculture* **22**(4): 79–83.

Faizool, S. and Ramjohn, R.K. (1995) *Agroforestry in the Caribbean*. Oficina Regional de la FAO para American Latina y El Caribe, Santiago, Chile, 35 pp.

FAO (1994) *Prácticas Agroforestales en el Departmento de Potosí Bolivia: Documemto de Trabajo No 1.* FAO/HOLANDA/CDF, Proyecto 'Desarrollo Forestal Comunal el Altiplano Boliviano', Potosí, Bolivia, 134 pp.

Feehan, J., Gillmor, D.A. and Culleton, N. (2005) Effect of an agri-environmental scheme on farmland biodiversity in Ireland. *Agriculture, Ecosystems and Environment* **107**(2–3): 276–286.

Finizio, A., Villa, S. and Vighi, M. (2005) Predicting pesticide mixtures load in surface water from a given crop. *Agriculture, Ecosystems and Environment* **111**(1–4): 111–118.

Fish, S.K. (2000) Hohokam impacts on Sonoran Desert environment, pp. 250–280. In: Lentz, D.L. (ed.), *Imperfect Balance Landscape Transformations in the Precolumbian Americas*. Columbia University Press, New York, 547 pp.

Fuente de la, M.A.S. and Marquis, R.J. (1999) The role of ant-tended extrafloral nectaries in the protection and benefit of a neotropical rainforest tree. *Oecologia* **118**(2): 192–202.

Fukai, S. (1993) Intercropping – bases of productivity. *Field Crops Research* **34**: 239–245.

Gamero, E.M., Lok, R. and Somarriba, E. (1996) Análisis agroecológico de huertos caseros traditionales en Nicaragua. *Agroforestería en las Americas* **3**(11/12): 36–40.

Garbaye, J., Delwaulle, J.C. and Duangna, D. (1988) Growth response in eucalyptus in the Congo to ectomycorrhizal inoculation. *Forest Ecology and Management* **24**: 151–157.

García-Barrios, L. and Ong, C.K. (2004) Ecological interactions, management lessons and design tools in tropical agroforestry systems. *Agroforestry Systems* **61–62**(1–3): 221–226.

Gent, F.W. (1681) *Systema Agriculturae: The Mystery of Husbandry Discovered*. Harrow, London, 134 pp.

Gillespie, A.R., Knudson, D.M. and Geilfur, F. (1993) The structure of four homegardens in the Petén, Gauatemala. *Agroforestry Systems* **24**: 157–170.

Glausiusz, J. (2003) A green renaissance for the Sahel. *Discover* **24**(1): 13.

Gliessman, S.R. (1998) *Agroecology: Ecological Processes in Sustainable Agriculture*. Ann Arbor Press, Chelsea, MI, 357 pp.

Glover, J. (1957) The relationship between total seasonal rainfall and the yield of maize. *Journal of Agricultural Science* **49**: 285–290.

Godoy, R. and Bennett, C.P.A. (1991) The economics of monocropping and intercropping by small holders: the case of coconuts in Indonesia. *Human Ecology* **19**(1): 83–98.

Goff, S.A. and Salmeron, J.M. (2004) Back to the future of cereals. *Scientific American* **291**(August – 2): 42–49.

Gold, C.S., Koggundu, A., Abera, A.M.K. and Karamura, D. (2002) Selection criteria of *Musa* cultivars through a farmer participatory appraisal survey in Uganda. *Experimental Agriculture* **38**(1): 29–38.

Gold, M.V. (1994) *Sustainable Agriculture: Definitions and Terms*. Special Reference SRB 94-05, National Agricultural Library, Beltsville, MD, 10 pp.

Gollin, D., Morris, M. and Beyerlee, D. (2005) Technology adoption in intensive post-green revolution systems. *American Journal of Agricultural Economics* **87**(5): 1310–1316.

Gomes de Almeida, S. and Fernandes, G.B. (2006) Economic benefits of a transition to ecological agriculture. *Leisa* **22**(2): 28–29.

Gongfu, Z. (1982) The mulberry dike–fish pond complex: a Chinese ecosystem of land–water interactions on the Pearl River Delta. *Human Ecology* **10**(2): 191–202.

Gonzáles, R.J. (2001) *Zapotec Science: Farming and Food in the Northern Sierra of Oaxaca*. University of Texas Press, Austin, TX, 328 pp.

Gosling, P. and Shepard, M. (2005) Long-term changes in soil fertility in organic arable farming systems in England, with particular reference to phosphorus and potassium. *Agriculture, Ecosystems and Environment* **105**(1–2): 425–432.

Gras, N.S.B. (1940) *A History of Agriculture in Europe and Americas.* F.S. Crofts and Co., New York, 496 pp.

Graves, A.R., Burgess, P.J., Liagre, F., Terreaux, J.-P. and Duprez, C. (2005) Development and use of a framework for characterizing computer models of silvoarable economics. *Agroforestry Systems* **65**(1): 53–65.

Green, S.B. (1908) *Principles of American Forestry*. John Wiley & Sons, New York, 334 pp.

Gregg, S.A. (1988) *Foragers and Farmers: Population Interaction and Agricultural Expansion in Prehistoric Europe.* University of Chicago Press, Chicago, 275 pp.

Gregory, M.M., Shea, K.L. and Bakko, E.B. (2005) Comparing agroecosystems: effects of cropping and tillage patterns on soil, water, energy use and productivity. *Renewable Agriculture and Food Systems* **20**(2): 81–90.

Grimm, V., Revilla, E., Berger, U., Jeltsch, F., Mooij, W.M., Railsback, S.F., Thulke, H.-H., Weiner, J., Wiegand, T. and DeAgelis, D.L. (2005) Pattern-oriented modeling of agent-based complex systems: lessons for ecology. *Science* **310**(11 November – 5750): 987–991.

Grodzinsky, A.M. (1992) Allelopathic effects of cruciferous plants in crop rotation, pp. 77–85. In: Rizvi, J.H. and Rizvi, V. (eds.), *Allelopathy: Basic and Applied Aspects*. Chapman and Hall, London, 490 pp.

Groppali, R. (1993) Breeding birds in traditional tree rows and hedges in the central Po valley (Province of Cremona, Northern Italy), pp. 153–158. In: Bunce, R.G.H., Ryszkowski, L. and Paoletti, M.G. (eds.), *Landscape Ecology and Agroecosystems*. Lewis Publishers, London, 241 pp.

Guillet, D. (2006) Rethinking irrigation efficiency: chain irrigation in Northwestern Spain. *Human Ecology* **34**(3): 305–329.

Gupta, G.N., Mutha, S. and Limba, N.K. (2000) Growth of *Albizia lebbeck* on micro-catchments in the Indian arid zone. *International Tree Crops Journal* **10**: 193–202.

Hansen, S.W. (1996) Is agricultural sustainability a useful concept? *Agricultural Systems* **50**: 117–143.

Harmsen, K. (2000) A modified Mitscherlich equation for rainfed crop production in semi-arid areas. 1. Theory. *Netherlands Journal of Agricultural Science* **48**: 237–250.

Harris, F.M.A. and Mohammed, S. (2003) Relying on nature: wild food in Northern Nigeria *Ambio* **32**(1): 24–29.

Harrison, G. (1775) *Agriculture Delineated*. J. Wilkie, London, 414 pp.

Hauser, S., Vanlauwe, B., Asawalam, D.O. and Norgrove, L. (1997) Role of earthworms in traditional and improved low-input agricultural systems in West Africa, pp. 113–136. In: Brussaard, L. and Ferrara-Cerrato, R. (eds.), *Soil Ecology in Sustainable Agricultural Systems*. Lewis Publishers, New York, 168 pp.

Herzog, F. and Oetmann, A. (2000) Communities of interest and agroecosystem restoration: Streuobst in Europe, pp. 85–102. In: Flora, C. (ed.), *Interactions Between Agroecosystems and Rural Communities*. CRC Press, Boca Raton, FL, 273 pp.

Huang, H.T. and Yang, P. (1987) The ancient cultured citrus ant. *BioScience* **37**: 665–667.

Hulugalle, N.R. and Ezumah, H.C. (1991) Effects of cassava-based cropping systems on physio-chemical properties of soil and earthworm casts in a tropical alfisol. *Agriculture, Ecosystems and Environment* **25**: 55–63.

Hulugalle, N.R. and Ndi, J.N. (1993) Effects of no-tillage and alley cropping on soil properties and crop yields in a typic kandivdult of southern Cameroon. *Agroforestry Systems* **22**: 207–220.

Husain, A.A. (2000) *Scent in the Islamic Garden: A Study of Deccani Urdu Literary Sources*. Oxford University Press, Oxford, UK, 284 pp.

IACPA (1998) *Integrated Farming: Agricultural Research into Practice.* A Report from the Integrated Arable Crop Production Alliance for Farmers, Agronomists and Advisors. Ministry of Agriculture, Fisheries and Food, MAFF Publications, London, 16 pp.

ICRAF (1993) *Annual Report*. International Centre for Research in Agroforestry, Nairobi, Kenya.

Ives, A.R. (2005) Empirically motivated ecological theory. *Ecology* **80**(12): 3137–3138.

Jashemski, W.F. (1987) Recently excavated gardens and cultivated land of the villas at Boscoreale and Oplontis, pp. 33–75. In: MacDougall, E.B. (ed.), *Ancient Roman Villa Gardens*. Dumbarton Oaks Research Library, Washington, DC, 260 pp.

Jensen, M. (1993) Soil conditions, vegetation structure and biomass of a Javanese homegarden. *Agroforestry Systems* **24**: 171–186.

Jobidon, R. (1992) Allelopathy in Quebec forestry – case studies in natural and managed ecosystems, pp. 341–356. In: Rizvi, J.H. and Rizvi, V. (eds.), *Allelopathy: Basic and Applied Aspects*. Chapman and Hall, London, 490 pp.

Jones, R.W. (2006) March of the weevils. *Natural History* **115**(February – 1): 30–35.

Jordan, C.F. (2004) Organic farming and agroforestry: alley cropping for mulch production for organic farms of southeastern United States. *Agroforestry Systems* **61–62**(1–3): 79–90.

Jordan, V.W.L., Hutcheon, J.A., Glen, D.M. and Farmer, D.P. (1996) *Technology Transfer of Integrated Farming Systems: The LIFE Project*, 3rd edition. Integrated Arable Crops Research, Bristol, UK, 24 pp.

Jordan, V.W.L., Hutcheon, J.A. and Donaldson, G.V. (1997a) The role of integrated arable production systems in reducing synthetic inputs. *Aspects of Applied Biology* **50**: 419–429.

Jordan, V.W.L., Hutcheon, J.A., Donaldson, G.V. and Farmer, D.P. (1997b) Research into and development of integrated farming systems for less-intensive arable crop production: experimental progress (1989–1994) and commercial implementation. *Agriculture, Ecosystems and Environment* **64**: 141–148.

Josiah, S.J., St-Pierre, R., Brott, H. and Brandle, J. (2004) Productive conservation: diversifying farm enterprises by producing specialty products in agroforestry systems. *Journal of Sustainable Agriculture* **23**(3): 93–108.

Kaiser, J. (2003) Sipping from the poisoned chalice. *Science* **302**(17 October – 5644): 376–379.

Kamara, A.Y., Akobundu, I.O., Chikoye, D. and Jutzi, S.C. (2000) Selective control of weeds in an arable crop by mulches from some multi-purpose trees in Southwestern Nigeria. *Agroforestry Systems* **50**: 17–26.

Kass, D.C.L. and Somarriba, E. (1999) Traditional fallows in Latin America. *Agroforestry Systems* **47**: 13–36.

Kaval, P. (2004) The profitability of alternative cropping systems: a review of the literature. *Journal of Sustainable Agriculture* **23**(3): 47–65.

Keating, B.A. and Carberry, P.S. (1993) Resource capture and use in intercropping: solar radiation. *Field Crops Research* **34**: 273–301.

Keil, A., Zeller, M. and Franzel, S. (2005) Improved tree fallows in smallholder maize production in Zambia: do testers adopt the technology. *Agroforestry Systems* **65**(3): 225–236.

Khan, Z.R., Pickett, J.A., van den Berg, J., Wadham, L.J. and Woodcock, C.M. (2000) Exploiting chemical ecology and species diversity: stem borer and striga control for maize and sorghum in Africa. *Pest Management Science* **56**: 957–962.

Kihanda, F.M., Warren, G.P. and Micheni, A.N. (2004) Effects of manure and fertilizers on grain yield, soil carbon, and phosphorus in a 13-year field trial in semi-arid Kenya. *Experimental Agriculture* **41**(4): 384–412.

Kirsch, P. (1988) Pheromones: their potential role in control of agricultural insect pests. *American Journal of Alternative Agriculture* **3**(2/3): 83–97.

Klironomos, J.N. (2002) Another form of bias in conservation research. *Science* **298**(25 October – 5594): 749.

Kumer, B.M. and Nair, P.K.R. (2004) The enigma of the tropical homegarden. *Agroforestry Systems* **61–62**(1–3): 135–152.

Kwesiga, F. and Coe, R. (1994) The effect of short rotation *Sesbania sesban* planted as fallows on maize. *Forest Ecology and Management* **64**: 199–209.

Lacy, S.M., Cleveland, D.A. and Soleri, D. (2006) Farmers choice of sorghum varieties in Southern Mali. *Human Ecology* **34**(3): 331–353.

Lagerlöf, J., Goffre, B. and Vincent, C. (2002) The importance of field boundaries for earthworm (Lumbricidae) in the Swedish agricultural landscape. *Agriculture, Ecosystems and Environment* **89**: 91–103.

Langyintuo, A.S. and Dogbe, W. (2005) Characterizing the constraints for the *Callopogonium mucunoides* improved fallow in rice production systems in northern Ghana. *Agriculture, Ecosystems and Environment* **110**(1–2): 79–90.

Lathrop, L.E. (1825) *The Farmer's Library.* William Fay, Rutland, VT, 215 pp.

Laurance, S.G.W. (2004) Landscape connectivity and biological corridors, pp. 50–64. In: Schroth, G., da Sonseca, G.A.B., Harvey, C.A., Gascon, C., Vasconcelos, H.L. and Izac, A.-M.N. (eds.), *Agroforestry and Biodiversity Conservation in Tropical Landscapes*. Island Press, Boca Ratio, FL, 523 pp.

Lelle, M.A. and Gold, M.A. (1994) Agroforestry systems for temperate climates: lessons from Roman Italy. *Forest and Conservation History* **38**: 118–126.

León, M.C. and Harvey, C.A. (2006) Live fences and landscape connectivity in a neotropical agricultural landscape. *Agroforestry Systems* **68**(1): 15–26.

Levins, R. and Vandermeer, J.H. (1990) The agroecosystem embedded in a complex ecological community, pp. 341–362. In: Carroll, C.R., Vandermeer, J.H. and Resset, P.R. (eds.), *Agroecology*. McGraw-Hill, New York, 641 pp.

Liebman, M. and Gallandt, E.R. (1997) Many little hammers: ecological management of crop–weed interactions, pp. 291–343. In: Jackson, L.E. (ed.), *Ecology in Agriculture*. Academic Press, New York, 474 pp.

Liebman, M., Mohler, C.L. and Staver, C.P. (2001) *Ecological Management of Agricultural Weeds.* Cambridge University Press, Cambridge, 532 pp.

Liebold, M.A. (1995) Mechanistic models and community context. *Ecology* **76**(5): 1371–1382.

Linares, O.F. (1976) Garden hunting in the American tropics. *Human Ecology* **4**(4): 331–349.

Liyanage, M. de S. (1993) The role of MPTS in coconut-based farming systems in Sri Lanka. *Agroforestry Today* **5**(3): 7–8.

Lotter, D.W., Seidel, R. and Liebhardt, W. (2003) The performance of organic and conventional cropping systems in an extreme climate year. *American Journal of Alternative Agriculture* **18**(3): 146–154.

Lowenberg-DeBoer, J., Krause, M., Deuson, R.R. and Reddy, K.C. (1990) *A Simulation of Millet and Cowpea Intercrop.* Station Bulletin 575, Agricultural Experiment Station, Purdue University, West Lafayette, IN, 82 pp.

Ma, M., Tarmi, S. and Helenius, J. (2002) Revisiting the species–area relationship in a semi-natural habitat: floral richness in agricultural buffer zones in Finland. *Agriculture, Ecosystems and Environment* **89**: 137–148.

MacDonald, J. (1994) *The Ornamental Kitchen Garden*. David and Charles, Devon, UK, 144 pp.

MacDonald, K.I. (1998) Rationality, representation, and the risk mediating characteristics of a Karakoram mountain farming system. *Human Ecology* **26**(2): 287–317.

MacKenze, D. (1999) Good ol' milkweed. *New Scientist* **164**(16 October – 2208): 24.

Mäder, P., Fliessbach, A., Dubois, D., Gunst, L., Fried, P. and Higgli, U. (2002) Soil fertility and biodiversity in organic farming. *Science* **296**(31 May – 5573): 1694–1697.

Makumba, W., Janssen, B., Oenema, O. and Akinnifesi, K. (2006) Influence of time of application on the performance of gliricidia prunings as a source of N for maize. *Experimental Agriculture* **42**(1): 51–63.

Mando, A. and Van Rheenen, T. (1998) Termites and agricultural production in the Sahel: from enemy to friend. *Netherlands Journal of Agricultural Science* **64**: 77–85.

Mann, C.C. (2005) *1491: New Revelations of the Americas Before Columbus.* Knopf, New York, 465 pp.

Marc, P., Canard, A. and Ysnel, F. (1999) Spiders (*Araneae*) useful for pest limitation and bioindication. *Agriculture, Ecosystems and Environment* **74**: 229–273.

Mas, A.H. and Dietsch, T.V. (2004) Linking shade coffee certification to biodiversity conservation: butterflies and birds in Chiapas, Mexico. *Ecological Applications* **14**(3): 642–654.

Matlack, G.R. (1994) Plant species migration in mixed-history forest landscapes in eastern North America. *Ecology* **75**(5): 1491–1502.

Matson, P.A., Parton, W.J., Power, A.G. and Swift, M.J. (1997) Agricultural intensification and ecosystem properties. *Science* **277**(25 July – 5325): 504–509.

Maudeley, M., Seeley, B. and Lewis, O. (2002) Spatial distribution patterns of predatory arthropods within an English hedgerow in early winter in relation to habitat variables. *Agriculture, Ecosystems and Environment* **89**: 77–89.

Mbabaliye, T. and Wojtkowski, P.A. (1994) Problems and perspectives on the use of a crop simulation model in an African research station. *Experimental Agriculture* **39**: 441–446.

McIntyre, R. (1999) Four-level forest farming operation making the most of agroforestry. *Inside Agroforestry* (Fall): 10.

McLean, T. (1980) *Medieval English Garden.* The Viking Press, New York, 298 pp.

McNeely, J.A. (1993) Economic incentives for conserving biodiversity: lesson from Africa. *Ambio* **22**(2–3): 144–150.

Mead, R. and Willey, J. (1980) The concept of 'land equivalent ratio' and the yield advantages from intercropping. *Experimental Agriculture* **16**: 217–228.

Méndez, V.E., Lok, R. and Somarriba, E. (2001) Interdisciplinary analysis of homegardens in Nicaragua: micro-zonation, plant use and socio-economic importance. *Agroforestry Systems* **51**: 85–96.

Menzel, P. and D'Alvisio, F. (1998) *Man Eating Bugs: The Art and Science of Eating Insects*. Ten Speed Press, Berkeley, CA, 191 pp.

Menzies, N. (1988) Three hundred years of taungya: a sustainable system of forestry in South China. *Human Ecology* **16**(4): 361–376.

Michon, G. and Mary, F. (1994) Conversion of traditional village gardens and new economic strategies of rural households in the area of Bogor, Indonesia. *Agroforestry Systems* **25**: 31–58.

Milius, S. (2003) Sweet lurkers. *Science News* **164**(13 December – 24): 374.

Millman, J. (2005) Companies seek out these tiny assassins to fight crop pests. *The Wall Street Journal* **CCXLVI**(18 November – 108): 1.

Molles, M.C. (1999) *Ecology: Concepts and Applications*. McGraw Hill, New York, 509 pp.

Montambault, J.R. and Alavalapati, J.R.R. (2005) Socioeconomic research in agroforestry: a decade in review. *Agroforestry Systems* **65**(2): 151–161.

Morales, H., Perfecto, I. and Ferguson, B. (2001) Traditional fertilization and its effect on corn insect populations in the Guatemalan highlands. *Agriculture, Ecosystems and Environment* **84**: 145–155.

Moynihan, E.B. (1979) *Paradise as a Garden in Persia and Mughad India*. George Braziller, New York, 168 pp.

Muschler, R.G. (2001) Shade improves coffee quality in a sub-optimal coffee-zone of Costa Rica. *Agroforestry Systems* **85**: 131–139.

Nair, P.K.R., Buresh, R.J., Mugendi, D.N. and Latt, C.R. (1999) Nutrient cycling in tropical agroforestry systems: myths and science, pp. 1–11. In: Buck, L.E., Lassoie, J.P.

and Fernandes E.C.M. (eds.), *Agroforestry in Sustainable Agricultural Systems*. CRC Press, New York, 416 pp.

Nandal, D.P.S. and Bisla, S.S. (1995) Growing oats under trees for more and better fodder. *Agroforestry Today* **7**(1): 12.

Nyakanda, C., Mashingaidze, A.B. and Zhakata, E. (1998) Early growth of pigeonpea under open-pit planting and restricted weeding in semi-arid Zimbabwe. *Agroforestry Systems* **41**: 267–276.

O'Brien, C. (2002) Bees buzz elephants. *The New Scientist* **176**(16 November – 2369): 8.

Ong, C. (1994) Alley cropping – ecological pie in the sky? *Agroforestry Today* **6**(3): 8–10.

Ong, C.K. (1991) Interactions of light, water and nutrients in agroforestry systems, pp. 107–121. In: Avery, M.E., Cannell, M.G.R. and Ong, C.K. (eds.), *Biophysical Research for Asian Agroforestry*. Winrock International, AK and Oxford and IBH Publishing Co., New Delhi, 285 pp.

Ong, C.K., Black, C.R., Marshall, F.M. and Corlett, J.E. (1996) Principles of resource capture and utilization of light and water, pp. 73–158. In: Ong, C.K. and Huxley, P. (eds.), *Tree–Crop Interactions: A Physiological Approach*. CAB International, Wallington, UK, 386 pp.

Palm, C.A. (1995) Contribution of agroforestry trees to nutrient requirements of intercropped plants. *Agroforestry Systems* **30**: 105–124.

Paris, Q. (1992) The von Liebig hypothesis. *American Journal of Agricultural Economics* **74**: 1019–1028.

Pattanayak, S. and Butry, D.T. (2005) Spatial complementarity of forests and farms: accounting for ecosystem services. *American Journal of Agricultural Economics* **87**(4): 995–1008.

Pearce, F. (2006) The miracle of stones. *New Scientist* **191**(2565): 50–51.

Pennisi, E. (2004) The secret life of fungi. *Science* **304**(11 June – 5677): 1620–1622.

Percival, J. (1922) *The Wheat Plant*. E.P. Dutton and Co., New York, 463 pp.

Perfecto, I. and Armbrecht, I. (2003) The coffee agroecosystem in the neotropics: combining ecological and economic goals, pp. 159–194. In: Vandermeer, J.H. (ed.), *Tropical Agroecosystems*. CRC Press, Boca Raton, FL, 268 pp.

Perfecto, I., Rice, R.A., Greenberg, R. and Van der Voort, M.E. (1996) Shade coffee: a disappearing refuge for biodiversity. *BioScience* **46**(8): 598–608.

Peterken, G.F. (1993) Long-term floristic development of woodland on former agricultural lands in Lincolnshire, England, pp. 31–43. In: Watkins, C. (ed.), *Ecological Effects of Afforestation: Studies in the History and Ecology of Afforestation in Western Europe*. CABI, New York, 224 pp.

Peters, W.J. and Neuenschwander, L.F. (1988) *Slash and Burn: Farming in the Third World*. University of Idaho Press, Moscow, ID, 113 pp.

Phillips, S.L. and Wolfe, M.S. (2005) Evolutionary plant breeding for low input systems. *Journal of Agricultural Science* **143**: 245–254.

Picone, C. (2003) Managing mycorrhizae for sustainable agriculture in the tropics, pp. 95–132. In: Vandermeer, J.H. (ed.), *Tropical Agroecosystems*. CRC Press, Boca Raton, FL, 268 pp.

Platt, J.O., Caldwell, J.S. and Kok, L.T. (1999) Effect of buckwheat as a flowering border on population of cucumber beetle and their natural enemies in cucumber and squash. *Crop Protection* **18**(5): 305–313.

Plucknett, D.L. and Smith, N.J.H. (1986) Historical perspectives on multiple cropping, pp. 20–39. In: Francis, C.A. (ed.), *Multiple Cropping Systems*. MacMillan Publishing Co., New York, 383 pp.

Poggio, S.L. (2005) Structure of weed communities occurring in monocultures and intercroping of field pea and barley. *Agriculture Ecosystems and Environment* **109**(1–2): 48–58.

Polak, P. (2005) The big potential of small farms. *Scientific American* **293**(September – 3): 84–91.

Pollan, M. The (agri)cultural contradictions of obesity. *The New York Times Magazine* (October 12): A10.

Prasad, R.B. and Brook, R.M. (2004) Effects of varying maize densities on intercropped maize and soybean. *Experimental Agriculture* **41**(3): 365–382.

Price, C. (1995) Economic evaluation of financial and non-financial costs and benefits in agroforestry development and the value of sustainability. *Agroforestry Systems* **30**: 75–86.

Prinsley, R.T. (1992) The role of trees in sustainable agriculture – an overview. *Agroforestry Systems* **20**: 87–115.

Racine, M., Boursier-Mougenot, E.J.-P. and Binet, F. (1987) *The Gardens of Provence and the French Riviera*. MIT Press, Cambridge, MA, 315 pp.

Raderama, S., Otieno, H., Atta-Krah, A.N. and Niang, A.I. (2004) System performance analysis of an alley-cropping system in western Kenya and its explanation by nutrient balances and uptake processes. *Agriculture Ecosystems and Environment* **104**(3): 631–652.

Ranganathan, R., Fafchamps, M. and Walker, T.S. (1991) Evaluating biological productivity in intercropping systems with production possibility curves. *Agricultural Systems* **36**: 137–157.

Rao, M.R. (1986) Cereals in multiple cropping, pp. 96–132. In: Francis, C.A. (ed.), *Multiple Cropping Systems*. MacMillian Publishing Co., New York, 383 pp.

Rao, M.R. and Gacheru, E. (1998) Prospects of agroforestry for *Striga* management. *Agroforestry Forum* **9**(2): 22–26.

Raynor, W. (1992) Economic analysis of indigenous agroforestry: a case study on Pohnpei Island, Federated State of Micronesia, pp. 243–258. In: Sullivan, G.H., Huke, S.M. and Fox, J.M. (eds.), *Financial and Economic Analysis of Agroforestry Systems: Proceedings of a Workshop Held in Honolulu, Hawaii, USA, July 1991*. Nitrogen Fixing Tree Association, Paia, HI, 312 pp.

Reynolds, H.L., Packer, A., Bever, J.D. and Clay, K. (2003) Grass roots ecology: plant–microbe–soil interactions as drivers of plant community structure and dynamics. *Ecology* **84**(9): 2281–2291.

Reynolds, L. (1991) Livestock in agroforestry: a farming systems approach, pp. 233–256. In: Avery, M.E., Cannell, M.G.R. and Ong, C.K. (eds.), *Biological Research for Asian Agroforestry*. Windrock International, USA and Oxford and IBH Publishing Co., New Delhi, 285 pp.

Rezaul Haq, A.H.M., Ghosal, T.K. and Ghosh, P. (2004) Cultivating wetlands in Bangladesh. *Leisa* **20**(4): 18–20.

Rice, R.A. and Greenburg, R. (2000) Cocoa cultivation and the conservation of biological diversity. *Ambio* **29**(3): 167–173.

Rigby, D. and Cáceres, D. (2001) Organic farming and the sustainability of agricultural systems. *Agricultural Systems* **68**: 21–40.

Roberts, P.A., Matthews Jr., W.C. and Ehlers, J.D. (2005) Root-knot nematode resistant cowpea cover crops in tomato production systems. *Agronomy Journal* **97**(6): 1626–1634.

Robinson, J. (2005) Silvopasture and the eastern wild turkey. *Agroforestry Notes* **28**(February): 1–4.

Sainju, U.M., Singh, B.P. and Whitehead, W.F. (2005) Tillage, cover crops, and nitrogen fertilization effects on cotton and sorghum root biomass, carbon, and nitrogen. *Agronomy Journal* **97**(5): 1279–1290.

Salafsky, N. (1994) Forest gardens in the Gunung Palung region of West Kalimantan, Indonesia. *Agroforestry Systems* **28**: 237–268.

Salam, A.M., Mohanakumaran, N., Jayachandran, B.K., Mammen, M.K., Sreekumar, D. and Babu, K.S. (1991) Thirty-one tree species support black pepper vines. *Agroforestry Today* **3**(4): 16.

Samita, S., Anputhas, M. and Abeysiriwardena, D.S. de Z. (2004) Accounting for multi traits in recommending rice varieties for diverse environments. *Experimental Agriculture* **41**(2): 213–225.

Sanchez, P.A. (1995) Science in agroforestry. *Agroforestry Systems* **30**: 5–55.

Sayre, L. (2005) Holding to the family farm. *Leisa* **21**(2): 16–17.

Schenck, C.A. (1904) *Forest Utilization, Mensuration and Silviculture*. Biltmore, North Carolina, 3 parts.

Scheu, S. and Parkinson, D. (1994) Effects of invasion of an aspen forest (Canada) by *Dendrobaena octaedra* (Lumbricidae) on plant growth. *Ecology* **75**(8): 2348–2361.

Schlich, W. (1910) *Schlich's Manual of Forestry*, Vol. II, 4th edition. Bradbusy, Agnew and Co., London, 424 pp.

Schroth, G. (1995) Tree root characteristics: as criteria for species selection and system design in agroforestry. *Agroforestry Systems* **30**: 125–143.

Schroth, G. (1999) A review of belowground interactions in agroforestry, focusing on mechanisms and management options. *Agroforestry Systems* **43**: 5–34.

Schulte, M. (1996) *Technología Agrícola Altoandina*. Swiss Agency for Development and Cooperation, 226 pp.

Schultz, B.B., Phillips, C., Rosset, P. and Vandermeer, J. (1982) An experiment in intercropping tomatoes and cucumber in southern Michigan. *Scientia Horticulturae* **18**: 1–8.

Setälä, H. (2000) Reciprocal interactions between Scots pine and soil web structure in the presence and absence of ectomycorrhiza. *Oecologia* **125**: 109–118.

Sharma, S.P. and Subehia, S.I.C. (2003) Effects of twenty-five years of fertilizer use on maize and wheat yields and quality of an acidic soil in the western Himalayas. *Experimental Agriculture* **39**(1): 55–64.

Shiva, V. (1991) *The Violence of the Green Revolution: Third World Agriculture, Ecology and Politics.* Zed Books, London, 263 pp.

Shrair, A.J. (2000) Agricultural intensity and its measure in frontier regions. *Agroforestry Systems* **49**: 301–318.

Shuster, W.D. and Edwards, C.A. (2003) Interactions between tillage and earthworms in agroecosystems, pp. 229–252. In: El Titi, A. (ed.), *Soil Tillage in Agroecosystems*. CRC Press, Boca Raton, FL, 367 pp.

Siame, J.A. (2006) The Mambwe mound cultivation system. *Leisa* **22**(4): 14–15.

Singleton, G.R., Sudarmaji, Jacob, J. and Krebs, C.J. (2005) Integrated management to reduce rodent damage to lowland crops in Indonesia. *Agriculture, Ecosystems and Environment* **107**(1): 75–82.

Slicher van Bath, B.H. (1963) *The Agrarian History of Western Europe, AD 500–1850.* Edward Arnold, Ltd., London, 346 pp.

Smith, M.E. and Francis, C.A. (1986) Breeding for multiple cropping systems, pp. 219–249. In: Francis, C.A. (ed.), *Multiple Cropping Systems*. MacMillan Publishing Co., London, 383 pp.

Smits, P.H. (1997) Insect pathogens: their suitability as biopecticides, pp. 21–28. In: *Microbial Insecticides: Novelty or Necessity? Proceedings of a Symposium organized by the British Crop Protection Council*, University of Warwick, Coventry, UK, 16–18 April, 301 pp.

Soemarwoto, O. (1987) Homegardens: a traditional agroforestry system with a promising future, pp. 156–170. In: Steppler, H.A. and Nair, N.P.K. (eds.), *Agroforestry: A Decade of Development*. ICRAF, Nairobi, Kenya, 335 pp.

Spitters, C.J.Y. and Kroff, M.J. (1989) *Modelling competition effects in intercropping systems.* Discussion paper for workshop on intercropping, IITA, Ibadan, Nigeria, 9–15 July, 19 pp.

Ståhl, L., Hogbery, P., Sellstedt, A. and Buresh, R.J. (2005) Measuring nitrogen fixation by *Sesbania sesban* planted fallows using 15N tracer techniques in Kenya. *Agroforestry Systems* **65**(1): 67–79.

Stakman, E.C., Bradfield, R. and Mangelsdorf, P.C. (1967) *Campaigns Against Hunger*. Harvard University Press, Cambridge, MA, 328 pp.

Stanton, M. and Young, T. (1999) Thorny relationships. *Natural History* **108**(9): 28–31.

Starcher A.M. (1995) *Good Bugs for Your Garden*. Algonquin Books, Chapel Hill, NC, 53 pp.

Stark, M., Mercado, A. and Garrity, D. (2000) Natural vegetative strip: farmers innovation gains popularity. *Agroforestry Today* **12**(1): 32–35.

Statälä, H. and Huhta, V. (1991) Soil fauna increase *Betula pendula* growth: laboratory experiments with coniferous forest floor. *Ecology* **72**(2): 665–671.

Steenbergh, W.F. and Lowe, C.H. (1969) Critical factors during the first year of life of the saguaro (*Cereus giganteus*) at Saguaro Monument, Arizona. *Ecology* **50**: 823–834.

Steuer, R.E. (1986) *Multiple Criteria Optimization: Theory, Computation, and Application*. John Wiley & Sons, Inc., New York, 546 pp.

Stockle, C.O., Papendick, R.I., Saxton, K.E., Cambell, G.S. and van Evert, F.K. (1994) A framework for evaluating the sustainability of agricultural production. *American Journal of Alternative Agricultural Production* **9**: 45–50.

Stocks, A. (1983) Candoshi and Cocamilla swiddens of eastern Peru. *Human Ecology* **11**(1): 69–84.

Suckling, D.M. and Karg, G. (1998) Pheromones and other semichemicals, pp. 63–99. In: Rechcigl, J.E. and Recheigl, N.A. (ed.), *Biological and Biotechnological Control of Insect Pests*. Lewis Publishers, Boca Raton, FL, 374 pp.

Swift, M.J., Izac, A.-M.N. and van Noordwijk, M. (2004) Biodiversity and ecosystem services in agricultural landscapes – are we asking the right questions? *Agriculture, Ecosystems and Environment* **104**(1): 113–134.

Swinkels, R. and Franzel, S. (1997) Adoption potential of hedgerow intercropping in maize-based cropping systems in the highlands of western Kenya. 2. Economic and farmers' evaluation. *Experimental Agriculture* **33**: 211–224.

Thornton, P.K., Dent, J.B. and Caldwell, R.M. (1990) Applications and issues on the modelling of intercropping systems in the tropics. *Agriculture, Ecosystems and Environment* **31**: 133–146.

Thurston, H.D. (1997) *Slash/Mulch Systems: Sustainable Methods for Tropical Agriculture*. Westview Press, Boulder, CO, 196 pp.

Tilman, D., Hill, J. and Lehman, C. (2006) Carbon-negative biofuels from low-input high-density grassland biomass. *Science* **314**(8 December – 5805): 1598–1600.

Tiwari, T.P., Brook, R.M. and Sinclair, F.L. (2004) Implications of hill farmers' agronomic practices in Nepal for crop improvement in maize. *Experimental Agriculture* **40**(4): 397–417.

Torquebiau, E. (1992) Are tropical agroforestry home gardens sustainable? *Agriculture, Ecosystems and Environment* **41**: 189–207.

van der Valk, H.C., Niassy, A. and BËye, A.B. (1999) Does grasshopper control create grasshopper problem? Monitoring side-effects of fenitrothion applications in the western Sahel. *Crop Protection* **18**(2): 139–149.

van der Vossen, H.A.M. (2004) A critical analysis of the agronomic and economic sustainability of organic coffee production. *Experimental Agriculture* **41**(4): 449–473.

van Noordwijk, M., Hairiah, K., Sitompul, S.M. and Syekhfani, M.S. (1992) Rotational hedgerow intercropping + *Peltophorum pterocarpum* = new hope for weed-infested soils. *Agroforestry Today* **4**(4): 4–6.

Vandermeer, J. (1989) *The Ecology of Intercropping*. Cambridge University Press, Cambridge, UK, 237 pp.

Vasey, D.E. (1992) *An Ecological History of Agriculture 10,000 BC–AD 10,000*. Iowa State University Press, Ames, IA, 363 pp.

Vaughan, M. and Black, S.H. (2006a) Agroforestry: sustaining native bee habitat for crop pollination. *Agroforestry Notes* AF(August)32. USDA, Lincoln, NE, 4 pp.

Vaughan, M. and Black, S.H. (2006b) Improving forage for native bee crop pollinators. *Agroforestry Notes* AF(August)33. USDA, Lincoln, NE, 4 pp.

Vera, F.W.M. (2000) *Grazing Ecology and Forest History*. CABI Publishing, The Netherlands, 506 pp.

Versteeg, M.N., Amadji, F., Eteka, A., Gogan, A. and Koudokpon, V. (1998) Farmers' adoptability of mucuna fallowing and agroforestry technologies in the coastal savanna of Benin. *Agricultural Systems* **56**: 269–287.

Vickery, J., Carter, N. and Fuller, R.J. (2002) The potential value of managed field margins as forging habitats for farmland birds in the UK. *Agriculture, Ecosystems and Environment* **89**: 41–52.

Waddell, E. (1975) How the Enga cope with frost: responses to climate perturbation in the Central Highlands of New Guinea. *Human Ecology* **3**(4): 249–273.

Walle, R.J. and Sims, B.G. (1998) Natural terrace formation through vegetative barriers on hillside farms in Honduras. *American Journal of Alternative Agriculture* **13**(2): 79–82.

Walter, G.H. (2003) *Insect Pest Management and Ecological Research*. Cambridge University Press, Cambridge, UK, 387 pp.

Wang, H. (1994) Tea and trees: a good blend from China. *Agroforestry Today* **6**(1): 6–8.

Wang, Q. and Shogren, J.F. (1992) Characteristics of the crop–paulownia system in China. *Agriculture, Ecosystems and Environment* **39**: 145–152.

Waring, R.H. and Pitman G.B. (1985) Modifying lodgepole pine stands to change susceptibility to mountain pine beetle attack. *Ecology* **66**(3): 884–897.

Watanabe, H. (1992) Tree–crop interactions in taungya plantations, pp. 32–43. In: Jordan, C.F., Gajaseni, J. and Watanabe, H. (eds.), *Taungya: Forest Plantations with Agriculture in Southeast Asia*. CAB International, Wallingford, UK, 153 pp.

Weaver, J.C. (1919) *The Ecological Relations of Roots*. Publication 286, Carnegie Institute of Washington, DC, 128 pp.

Weber, F.R. (1986) *Reforestation in Arid Lands*. Volunteers in Technical Assistance, Arlington, VA, 335 pp.

Weinzierl, R.A. (1998) Botanical insecticides: soaps and oils, pp. 101–121. In: Rechcigl, J.E. and Rechcigl, N.A. (eds.), *Biological and Biotechnological Control of Insect Pests*. Lewis Publishers, Boca Raton, FL, 374 pp.

Weiss, E., Kislev, M. and Hartmann, A. (2006) Autonomous cultivation before domestication. *Science* **312**(16 June – 5780): 1608–1610.

Wijesinghe, D.K. and Hutchings, M.J. (1999) The effects of environmental heterogeneity on the performance of *Glechoma hederacea*: the interactions between patch contrast and patch scale. *Journal of Ecology* **87**: 860–872.

Williams, T. (2002) America's largest weed. *Audubon* **104**(1): 24–31.

Wilson, J. (1994) *Landscaping with Herbs*. Houghton Mifflin Company, New York, 220 pp.

Wilson, P., Gibbons, J. and Ramsden, S. (2003) The impact of cereal prices and policy on crop rotation and supply response. *Journal of Agricultural Economics* **54**(2): 313–323.

Wilson, W.M. (2003) Cassava (*Manihot esculenta* Crantz), cyanogenic potential, and predation in northwestern Amazonia: the Tukanoan perspective. *Human Ecology* **31**(3): 403–416.

Wojtkowski, P.A. (1993) Toward an understanding of tropical home gardens. *Agroforestry Systems* **24**: 215–222.

Wojtkowski, P.A. (1998) *The Theory and Practice of Agroforestry Design*. Science Publishers, Enfield, NH, 282 pp.

Wojtkowski, P.A. (2002) *Agroecological Perspectives in Agronomy, Forestry, and Agroforestry.* Science Publishers, Enfield, NH, 356 pp.

Wojtkowski, P.A. (2004) *Landscape Agroecology*. Haworth Press, Binghamton, NY, 330 pp.

Wojtkowski, P.A. (2006a) *Introduction to Agroecology: Principles and Practices*. Haworth Press, Binghamton, NY, 375 pp.

Wojtkowski, P.A. (2006b) *Undoing the Damage: Silviculture for Ecologists and Environmental Scientists.* Science Publishers, Enfield, NH, 313 pp.

Wojtkowski, P.A., Brister, G.H. and Cubbage, F.W. (1988) Using multiple objective programming to evaluate multi-participant agroforestry systems. *Agroforestry Systems* **7**: 185–195.

Wojtkowski, P.A., Jordan, C.F. and Cubbage, F.W. (1991) Bioeconomic modeling in agroforestry: a rubber–cacao example. *Agroforestry Systems* **14**: 163–177.

Worlidge, J. (1669) *Systema Agriculturae.* F.W. Gent, London, 285 pp.

Yin, R. and He, Q. (1997) The spatial and temporal effects of paulownia intercropping. *Agroforestry Systems* **37**: 91–109.

Yoon, C.K. (2000) Simple method found to increase crop yield vastly. *The New York Times* (22 August): D1–D2.

Yoshida, T. and Kamitani, T. (1997) The stand dynamics of a mixed coppice forest of shade tolerant and intermediate species. *Forest Ecology and Management* **95**: 35–43.

Zancan, S., Trevisan, R. and Paoletti, M.G. (2006) Soil algae composition under different agroecosystems in north-eastern Italy. *Agriculture, Ecosystems and Environment* **112**(1): 1–12.

Zhang Fend (1996) Influences of shelterbelts in Chifeng, Inner Mongolia. *Unasylva* **47**(185): 11–15.

Zimmerer, K.S. (1999) Overlapping patchwork of mountain agriculture in Peru and Bolivia toward a regional-global landscape model. *Human Ecology* **27**(1): 135–165.

Index